Carbon Politics and the Failure of the Kyoto Protocol

"This book clarifies the complexities of international climate change negotiations. With in-depth understanding of the Carbon Game the policy makers will more likely come to the agreement in the post-Kyoto era." – Dr Wojciech Budzianowski, Professor at Wrocław University of Technology, Poland.

Carbon Politics and the Failure of the Kyoto Protocol charts the framework and political evolution of the Kyoto Protocol negotiations and examines the ensuing failure of the international community to adequately address climate change. The focus is not on the science or consequences of climate change but on the political gamesmanship of the major players throughout the United Nations Framework Convention on Climate Change negotiation process. More than an updated history of the subject matter, this book provides a detailed study of the carbon targets which became the biggest influencing factor on the reaction of nations to Kyoto's binding agreements.

The book provides an in-depth analysis of the leading nations' motives, including the US, China and Germany, in entering the negotiations; in particular, their economic interests. Despite the effort to combat climate change in politics that the negotiations represent, the book concludes that an agreement which requires almost 200 very different nations to agree on a single protocol is doomed to failure. The book offers a novel contribution to our understanding of this failure and suggests alternative frameworks and policies to tackle what is arguably the most complex political issue of our time.

Gerald Kutney is Managing Director of Sixth Element Sustainable Management, a renewable energy technologies consultancy, and was until recently Adjunct Professor of Environmental Sciences at the University of Northern British Columbia, Canada.

Routledge Explorations in Environmental Studies

Nuclear Waste Management and Legitimacy
Nihilism and responsibility
Mats Andrén

Nuclear Power, Economic Development Discourse and the Environment
The case of India
Manu V. Mathai

Federalism of Wetlands
Ryan W. Taylor

Governing Sustainable Urban Renewal
Partnerships in action
Rory Shand

Employee Engagement with Sustainable Business
How to change the world whilst keeping your day job
Nadine Exter

The Political Economy of Global Warming
The terminal crisis
Del Weston

Communicating Environmental Patriotism
A rhetorical history of the American environmental movement
Anne Marie Todd

Environmental Justice in Developing Countries
Perspectives from Africa and Asia-Pacific
Rhuks Temitope Ako

Carbon Politics and the Failure of the Kyoto Protocol

Gerald Kutney

Routledge
Taylor & Francis Group

LONDON AND NEW YORK

First published 2014
by Routledge
2 Park Square, Milton Park, Abingdon, Oxfordshire OX14 4RN

and by Routledge
711 Third Avenue, New York, NY 10017

Routledge is an imprint of the Taylor and Francis Group, an informa business

First issued in paperback 2015

© 2014 Gerald Kutney

The right of Gerald Kutney to be identified as author of this work has been asserted by him in accordance with sections 77 and 78 of the Copyright, Designs and Patents Act 1988.

British Library Cataloguing in Publication Data
A catalogue record for this book is available from the British Library

Library of Congress Cataloging in Publication data
Kutney, Gerald, 1953–
Carbon politics and the failure of the Kyoto Protocol / Gerald Kutney.
   pages cm. – (Routledge explorations in environmental studies)
ISBN 978-0-415-72931-4 (hardback) – ISBN 978-1-315-85109-9 (ebk)
1. United Nations Framework Convention on Climate Change (1992). Protocols, etc., 1997 Dec. 11.
2. Carbon dioxide mitigation–Law and legislation.   3. Environmental law, International.   I. Title.
K3593.A41992K88   2014
344.04′633–dc23
2013026667

ISBN 978-0-415-72931-4 (hbk)
ISBN 978-1-138-18688-0 (pbk)
ISBN 978-1-315-85109-9 (ebk)

Typeset in Times by
Out of House Publishing

To my dear wife Denise, who patiently proofread the manuscript with great skill and dedication.

Contents

Figures

Tables

Preface

Our species has been defined by the use of tools. A more fitting distinction is that we are the only species to abuse our environment with the use of tools. Yesterday's tools have mutated into today's engines, which convert the Earth's resources into a burden by releasing unnatural discharges in unnatural quantities as we create national capital at the expense of natural capital. And the release of one material in particular dwarfs all others, which just happens to directly come from the engines themselves. Anthropogenic carbon dioxide (CO_2) and other greenhouse gas (GHG) emissions, especially from electricity generation and the driving of vehicles, are at a rate of 1,000 tonnes of CO_2 every second. We have entered the "carbon century," the moniker for the twenty-first century. The scientific evidence is clear that our consumption of fossil fuels has begun to change the planet's climate. Our civilization depends on these engines but, ironically, the production of energy also threatens our civilization. We understand that we cannot continue on the same path as in the past, but we cannot simply pull the plug or turn off the switch either. There is an inseparable bond between carbon emissions and economic development. If we look at the US, China and the European Union (EU), this group accounted for 61 percent of GDP and 56 percent of CO_2 emissions in the world. The similarity is no mere coincidence, as economic prosperity and carbon emissions are both related to energy. Our civilization is carbonaceous.

The governments of the world have united in this cause under the United Nations Framework Convention on Climate Change (UNFCCC) and the Kyoto Protocol. As the first phase of Kyoto has ended, the time is upon us to reflect on what has been done. While we may commemorate such anniversaries, the results of our global efforts are not cause for celebration. Emissions are higher now than ever before!

Carbon Politics and the Failure of the Kyoto Protocol charts the framework and political evolution of the Kyoto negotiations in search for an answer to the international community's failure to act on climate change. The focus is not on the science or consequences of climate change but on the political gamesmanship which has been pervasive throughout the Kyoto negotiations by the major players. The influence of politics on the Kyoto process has been well studied,[1] and *Carbon Politics* is more than an updated history of the subject matter. What is new is the detailed study of the Kyoto targets which became, arguably, the major influence on the reaction of nations to Kyoto. An objective of the book is to identify what went wrong with Kyoto and to clear up popular misconceptions on global leadership on climate change legislation. Without such knowledge, a way forward to reduce global emissions cannot be achieved. At the end of *Carbon Politics*, framework and policy suggestions are made to lower emissions going forward.

Carbon Politics begins with the study of the history of climate science and the reaction, or lack thereof, of early government administrations. As external pressure grew for action, governments sent the ball back to the scientific court by demanding a consensus of the problem. A special organization to undertake this mission was created by the UN, the Intergovernmental Panel on Climate Change (IPCC), which was the official source of climate science for the Kyoto negotiators. The evolution and inner workings of this critical group are examined, and an answer is provided on why the powerful science from the IPCC has not done more to motivate governments to reduce emissions.

Alleged scandals, colloquially known as "climategate" and "glaciergate," rocked the foundations of both the IPCC and the climate change negotiations, and crippled the important conference at Copenhagen. These incidents set into motion a series of national inquiries into climate science that crossed two continents in 2010, which was an especially bad year for climate negotiations; the factors that caused this *annus horribilis* are analyzed. A root cause analysis of glaciergate is presented along with an overview of the response of the IPCC to the opportunities that this created. And an evaluation of climategate revealed how the Freedom of Information Act could be a threat to science when abused. Intimately involved in these scandals were the climate skeptics. In particular, those with political power have harassed and threatened some of the leading climate scholars. The political climate skeptic movement is a travesty of justice against the scientific community and is unprecedented in a democracy.

Next in *Carbon Politics*, we examine the actions and motivations of the major players concerning climate change regulations. With China, the analysis presents an explanation of their position, which is based on the interpretation of the pivotal phrase in the negotiations: "common but differentiated responsibilities"; this phrase has been intensely debated for over two decades without resolution, and is highlighted in *Carbon Politics*. Canada is unique among the members of Kyoto as it is the only country in the world to have withdrawn from it after ratifying the treaty. An analysis demonstrates that the Kyoto targets had placed an inequitable burden on the nation. The issues with the US and EU are more intriguing. Market mechanisms drove American strategy to the detriment of more important issues; the story on how this came about is studied in detail. We must conclude that climate change negotiation by the US is one of the great failures of international treaty discussions in modern times. They had been outmaneuvered by their European counterparts in a manner that would have impressed Machiavelli.

The media and scholarly literature imply that there is an "attitude" problem with some nations, especially the US, with regard to climate change legislation and carbon targets. Looking into this issue, I found reports that ascribed sociological and psychological barriers,[2] ideological differences between conservatives and liberals, and political[3] differences, to explain national variation. My analysis of global and national GHG emissions suggested that there really was not much difference between these nations: the dominant macro-factor that influenced their political response was a more fundamental principle, namely economics. Of course, there are reports documenting the role of economic factors. However, the bulk of these studies have focused on the United States and have created the false impression, even among global experts,[4] that countries with meaningful emissions standards, for example Germany, are willing to accept economic harm for the sake of the planet. The provocative findings of this work, on the other hand, are that economic drivers have determined the policies of all nations. Even those nations who advocated strict emissions standards demonstrated little concern for the actual environment in their decision-making process. In particular, the single most important influence on the implementation of carbon targets was found to be international competitiveness, which was more influenced by the direction that national emissions were headed than the absolute emissions themselves. The Kyoto targets evolved from politics not science. Those nations with already declining emissions became strong advocates for aggressive GHG emissions targets for *all* nations. The Kyoto Protocol became a tool to formally exploit nations with growing emissions by those with shrinking emissions, under the guise of a mechanism to address climate change.

The evidence indicates that a fundamental cause of the failure of the Kyoto negotiations was the inequitable carbon targets. Contrary to the absolute value of the reduction in carbon emissions, the Kyoto targets of the EU and even Germany were far easier to achieve than those of the US (and Canada). The EU had taken advantage of their position to improve their international competitiveness by forcing other developed nations to take on the burden of similar targets, even though the Europeans could just follow business-as-usual approaches to reach these goals. And no one wants to play a game when the rules favor their opponents.

While reaching such global agreements are immense bureaucratic accomplishments, the results must speak for themselves. The failure to even slow the tide of GHG emissions is the ultimate judgment on the Kyoto Protocol. Going forward will call for a new game plan. In the final chapter, a strategy to win the carbon game is proposed. A straightforward scheme is offered which some may criticize for being too simple, but which I believe will be more effective than the Kyoto Protocol.

The intended audience of this work is the educated reader, but would be of interest to climate change policymakers, academia and other scholars. The manuscript includes dozens of peer-reviewed references;[5] that said, the manuscript is not a critique of the literature on the failure of Kyoto. My initial analysis had been on the GHG emissions data from the period; it immediately became apparent that the Kyoto targets favored some nations. Further study reinforced this conclusion, yet such fundamental data and the political implications had been largely ignored in the scholarly literature. *Carbon Politics*, then, is an independent work on the failure of Kyoto based on the GHG emissions data, supported by the public record.

Such international negotiations are intense and daunting, far beyond which an outsider, such as me, can ever appreciate. The author would like to thank all those with much more direct experience that provided information, criticisms and comments, including: Palava Bagla, Jim Bruce, Liz Dowdeswell, Kimo Goree, Iqbal Hasnain, Georg Kaser, Ross McKitrick, Rafe Pomerance, Lisa Schipper, and John Stone.

References[6]

American Psychological Association, Task Force on the Interface Between *Psychology and Global Climate Change*, Psychology & Global Climate Change, March 2010; http://www.apa.org/science/about/publications/climate-change.aspx

Bolin, B., *A History of the Science and Politics of Climate Change*, Cambridge University Press, 2007

Boykoff, M.T. (ed.), *The Politics of Climate Change, A Survey*, Routledge, 2010

Brulle, R.J., Carmichael, J., and Jenkins, J.C., Shifting Public Opinion on Climate Change, *Climatic Change*, February 2, 2012

Cass, L.R., *The Failures of American and European Climate Policy*, SUNY, 2006

Dessler, A., and Parson, E.A., *The Science and Politics of Global Climate Change: a Guide to the Debate*, Cambridge University Press, 2006

Gaan, N., *Climate Change and International Politics*, Kalpaz, 2008

Goldenberg, S., Obama Environment Agenda Under Threat from Incoming Republicans, *Guardian*, October 31, 2010

Harrison, K., The Road Not Taken: Climate Change Policy in Canada and the United States, *Global Environmental Politics*, 7:4, p. 99, November 2007

Helm, D., and Hepburn, C. (eds.), *The Economics and Politics of Climate Change*, Oxford University Press, 2009

Leiserowitz, A., Mailbach, E., Roser-Renouf, C., and Smith, N., Global Warmings Six Americas, *Yale Project on Climate Change Communications*, May 2011; http://environment.yale.edu/climate/publications/SixAmericas May2011

Marsden, W., *Fool's Rule, Inside the Failed Politics of Climate Change*, Vintage Canada, 2012

Massai, L, *The Kyoto Protocol in the EU*, Springer, 2011

McCright, A.M., and Dunlap, R.E., The Politicization of Climate Change and Polarization in the American Public's View of Global Warming, 2001–2010, *Sociological Quarterly* 52 (2), 155, 2011

Nisbet, M.C., Communicating Climate Change, *Environment* 51 (2), 14, 2009

Perlmutter, D.D., and Rothstein, A.L., *The Challenge of Climate Change: Which Way Now?*, Wiley, 2011

Pooley, E., *The Climate War*, Hyperion, 2010

Scaglia, B., *The Politics of Global Warming*, Webster, 2011

Schneider, S., *Science as a Contact Sport, Inside the Battle to Save Earth's Climate*, National Geographical Society, 2009

Simpson, J., Jaccard, M., and Rivers, N., *Hot Air, Meeting Canada's Climate Change Challenge*, Emblem, 2007

Smith, N., and Leiserowitz, A., The Rise of Global Warming Skepticism: Exploring Affective Image Associations in the United States over Time, *Risk Analysis* (online), April 4, 2012; http://onlinelibrary.wiley.com/doi/10.1111/j.1539-6924.2012.01801.x/full

Spence, A., Poortinga, W., and Pidgeon, W., The Psychological Distance of Climate Change, *Risk Analysis* (online), October 12, 2011; http://onlinelibrary.wiley.com/doi/10.1111/j.1539-6924.2011.01695.x/full

Weber, E.U. and Stern, P.C., Public Understanding of Climate Change in the United States, *American Psychologist*, 66 (4), 315, 2011

Zia, A., *Post-Kyoto Climate Governance*, Routledge, 2013

Notes

1 For examples of recent works see Cass, 2006; Dessler and Parson, 2006; Bolin, 2007; Harrison, 2007; Simpson *et al.*, 2007; Gaan, 2008; Helm and Hepburn (eds.), 2009; Schneider, 2009; Pooley, 2010; Boykoff, M.T. (ed.), 2010; Perlmutter and Rothstein, 2011; Scaglia, 2011; Massai, 2011; Marsden, 2012; Zia, 2013.

2 American Psychological Association, pp. 8, 68; in 2011, an entire edition (May–June) of *American Psychology*, entitled *Psychology and Global Climate Change*, was based on this topic; also see Leiserowitz *et al.*, May, 2011; Spence *et al.*, October 12, 2011.

3 Leiserowitz, *et al.*, 2011; Nisbet, 2009; Smith and Leiserowitz, April 4, 2012; Weber and Stern, 2011; Goldenberg, October 31, 2010; McCright and Dunlap, 2011; Brulle *et al.*, February 2, 2012.

4 See, for example, Bolin, 2007, p. 181; Schneider, 2009, p. 156.

5 Five hundred or so references, a combination of scholarly works, news reports and public documents, were utilized in this study.

6 All web-based references were confirmed in June 2013, unless otherwise stated.

1   The players' association
The United Nations

United Nations' treaties are forged in their own world, one that many of us know very little about. Debates over a phrase can extend over years, and an agreement on a single paragraph can be hailed as an important breakthrough. Within this daunting and drawn out process, the treaty itself can unintentionally become the goal. Kyoto fell into this trap, as the original intent to reduce GHG emissions was sacrificed for the sake of reaching a deal, and the practical result is that the Protocol has become little more than a reporting mechanism. What good is a global climate treaty where most nations have no restraints on their GHG emissions? So how did this happen? We begin by examining the framework of the sensitive and complex negotiations leading to the Kyoto Protocol.

The road to Villach & Toronto

The general issue of borderless pollution problems had been first addressed by the United Nations at the Conference of the Human Environment in Stockholm[1] back in 1972. Upon a recommendation from the Stockholm meeting, the United Nations Environment Programme (UNEP) was formed on September 2, 1973. Maurice Strong (1929–)[2] of Canada, the Secretary-General of the conference, was appointed the first head of UNEP. He was followed by Mostafa Kamal Tolba (1922–) and then another Canadian, Elizabeth Dowdeswell,[3] who had been a member of the IPCC and Intergovernmental Negotiating Committee (INC).

UNEP adopted the framework convention[4] for international treaty negotiations. I will briefly review the structure of this type of treaty mechanism for those unfamiliar with the jargon, in order to better understand the Kyoto negotiations. The framework convention is a two-stage, multi-year process:

1. Framework Convention – a preliminary treaty, containing a general system of governance, institutional mechanisms and policy guidelines (i.e., the "framework") describing intent but few obligations (similar to a memorandum of understanding in the business world).
2. Protocol – a modification of the framework convention treaty containing specific legally binding commitments (similar to a contract in the business world). Protocols are often named after the city where agreement was reached.

Each stage of the treaty follows three steps:

1. negotiated agreement
2. national signature (largely symbolic)
3. national ratification (makes it binding).

Actual meetings where the treaty terms are negotiated are the sessions of the:

1. Intergovernmental Negotiating Committee (INC) for the preliminary framework convention, which usually disbands after the treaty has been signed.
2. Conference of the Parties (COP) for the protocol acceptance and other modifications to the framework convention.
3. Meeting of the Parties (MOP) is responsible for the protocol. In some cases, the COP also acts as the MOP, and is abbreviated as CMP.[5]

Both the INC and COP are supported by Working Groups and/or *Ad Hoc* Working Groups; the latter is a "temporary" group dealing with a specific task, especially challenging policies. Draft documents from the Working Groups incorporate "brackets" of objected terms and phrases, which allows the discussion to move forward without resolving all contentious clauses. During the Kyoto negotiations, the process degenerated, and the documents became almost comical; drafts were often nothing but a series of brackets, and brackets within brackets within brackets; the originator of a bracket was not identified, and countries often forgot who had objected to what.

After a specified majority of nations have signed on to the framework convention, it is officially in force, and negotiations can begin on amendments of legally binding commitments, called the protocols. The Conference of the Parties (COP) takes over negotiations from the INC at this stage, and the COP is the supreme decision-making body for any modifications to the framework convention. The COP meeting is divided into two sessions. In the first session, bureaucrats hash through the endless administrative duties and develop critical policies. In the second part of the Conference, the high-level meetings take place, which often includes senior ministers and heads of state. Upon reaching the threshold number of signatures by member states, the protocol becomes incorporated into the treaty, which is still called the framework convention. There is no legal obligation until the protocol has been "ratified" by the individual national governments.

The first framework convention by UNEP took place in 1976 at the Barcelona Convention for the Protection of the Mediterranean Sea against Pollution. Another early framework convention on the environment was in 1979, the Convention on Long-Range Transboundary Air Pollution (LRTAP;[6] in 1999, the Gothenburg Protocol was adopted as part of the LRTAP). The treaty to deal with acid rain had been organized through the United Nations Economic Commission for Europe.[7]

Acid rain & the flexible mechanisms

Despite the LRTAP, legislation was slow to develop in Europe. The turning point was when Germany, which had resisted any acid rain legislation, discovered in 1982 that acid rain was causing widespread damage to its forests.[8] In December 1983, the European Community issued a program to curtail emissions, called the Large Combustion Plant Directive (LCPD). The negotiations over LCPD were typical for such treaties. Among the member states, their political position reflected the impact such legislation would have on their economies.[9] The strongest objections came from the UK, which had the largest emissions; they wanted scientific proof before taking action.[10] There was a North–South split, with the poorer southern nations in Europe arguing that the richer countries should tackle most of the problem. After five years of debate, a watered-down agreement was reached allowing most states to continue a business-as-usual approach, with differentiated targets for each member.[11] Incentives in the UK were provided for coal-fired utilities to switch to natural gas.[12] An unintentional by-product of the LCPD was that the UK also began lowering its CO_2 emissions well before climate change was on the political agenda.

Acid rain was also a problem in the United States, where a creative approach was adopted: namely a cap and trade system and the related "flexible mechanisms" of emissions trading. The concept of a market approach to pollution control had been proposed by Ronald Harry Coase (1910–2013) in 1960, and six years later, Thomas D. Crocker had suggested a cap and trade scheme.[13] The free-market concept to control emissions was outlined independently in greater detail in 1968 by John Harkness Dales[14] (1920–2007), an economist at the University of Toronto, in his book *Pollution, Property & Prices*. His basic premise was that a cost–benefit analysis could be applied to pollution regulations.[15] The traditional policy of polluters simply reducing their emissions using the least expensive technology to meet regulations was replaced with a market-based system, in which regulated companies could pool their resources; a firm had the option to buy pollution credits from other regulated companies that were able to achieve greater emissions reductions than required. In other words, it was possible for individual polluters to comply with the regulations without actually reducing their emissions as long as the overall emissions met targets.

The Crocker–Dales model was put into practice by the 1990 Clean Air Act in the US. The proposal, called "Project 88," had been designed by Senators John Heinz (1938–91) and Tim Wirth (1939–).[16] The target had been set to reduce emissions to 8.1 Tg[17] by 2010. That goal was reached in 2007, and in 2009 emissions were down to 5.7 Tg.[18] This free-market environmental program was hailed as a great triumph. But there were some skeptics. Congressman Tom Coburn (1948–) expressed doubts about the market-based mechanism as the main reason for success had been the availability of low-sulfur coal from Wyoming.[19]

In the end, the acid rain problem in Britain and the US was largely resolved more by economics than any special policy initiatives, as low-priced natural gas was available in the UK, and low-priced low-sulfur coal from Wyoming. In other words, the market itself moved industry towards low SO_2 emissions, although legislation had encouraged the change. Similar success has not been repeated with CO_2 emissions even though the Crocker–Dales model is part of the Kyoto Protocol. Ominously, Crocker and Dales themselves had questioned whether their mechanism would work to combat climate change because of the greater complexities involved.[20]

The ozone hole & the Montreal protocol

The first atmospheric problem that transcended oceans as well as borders was the ozone hole. Whereas CO_2 absorbs infrared radiation, ozone absorbs ultraviolet radiation. By doing so, ozone protects us from harmful UVB radiation, a contributor to skin cancer. Science had found that the protective ozone layer was being depleted by a group of commonly used chemicals called chlorofluorocarbons (CFCs). In 1976, the National Research Council issued a warning about the declining levels of ozone.[21] And five years later, UNEP formed the "*Ad Hoc* Working Group of Legal and Technical Experts for the Elaboration of a Global Framework for the Protection of the Ozone Layer." These discussions led to the Vienna Convention for the Protection of the Ozone Layer four years later, and a legally binding protocol was reached in Montreal in 1987.[22]

Among global agreements, the ozone negotiations progressed at a record pace. The US took the international lead while the EU (European Union)[23] attempted to derail the discussions (when it came to climate change, though, the roles were reversed). The US and the UK were both major producers of CFCs. However, the dominant US producer, DuPont, had discovered a substitute,[24] so any protocols would be a competitive advantage to the largest American producer of CFCs.

The acid rain and ozone negotiations illustrate a common feature of such negotiations. The environment had been the reason for nations to come together, but once at the negotiating table, economic interests take over. The same fate would befall climate change. Tackling climate change, though, is a much more complex matter than acid rain or "the ozone hole."[25]

Villach & Toronto

The notion that relatively trivial changes to the atmosphere could impact our climate is an extraordinary one; the famous Swedish chemist, Svante Arrhenius (1859–1927), was the first to make this counter-intuitive connection. Following the fundamental experiments of John Tyndall (1820–93) on greenhouse gases, Arrhenius deduced that CO_2 could influence world temperatures and that emissions from the burning of fossil fuels would increase the world's temperature.[26] Arrhenius continued his studies, predicting in 1906 that a doubling of CO_2 levels would cause global temperatures to rise by 1.6°C. But the Swedish chemist did not view climate change as a threat; a warmer climate would allow more crops to grow.[27]

Arrhenius' papers created little interest within the general scientific community. Had it not been for Guy Stewart Callendar (1898–1964), the field of study may have fallen into obscurity. In his first paper on the subject in 1938, "The Artificial Production of Carbon Dioxide and Its Influence on Temperature," he calculated that a doubling of CO_2 would increase temperatures by 2°C. The following year he warned of the "grand experiment" that was an "agent of global change." Through the 1930s and 1940s, Callendar was largely alone in the fledgling field of climate change. In one of his last papers, in 1958, he showed a graph of increasing CO_2 levels in the atmosphere from 1870–1955.[28]

Since then, the science of climate change has progressed through three major periods:[29]

1. *1950–84* (the Era of Knowledge Building) – Within the confines of physical science journals and conferences, a general consensus emerged among the scientific community about climate change caused by anthropogenic carbon dioxide emissions.
2. *1985–97* (the Pre-Kyoto Era) – The international community began to take notice with the Villach Conference in 1985, the *Brundtland Report* in 1987 and the Toronto Conference in 1988. The IPCC was formed by the UN (1988), and issued the First Assessment Report (1990) and the Second Assessment Report (1995). Political leaders responded at the Earth Summit with the UNFCCC in 1992. The serious negotiations then began, and the Kyoto Protocol was signed in 1997.
3. *1998–2012* (the Kyoto Era) – 18 COP and eight CMP meetings were held, and two more Assessment Reports were issued by the IPCC. The first decade of the new millennium witnessed far greater global GHG emissions than ever before.

During the 1950s the scientific study of climate change came into its own, with Canadian-born Gilbert Plass (1920–2004) playing a pivotal role.[30] He predicted in 1956 that CO_2 levels would increase by 30 percent and the temperature by 1°C by 2000, compared to levels at the beginning of the century (actual results for the century were an increase in CO_2 levels of 30 percent and an increase in temperature of 0.8°C). In the same year, Roger Revelle (1909–91) introduced climate change to the general public through an interview in *Time*; the article began:[31]

> Since the start of the industrial revolution, mankind has been burning fossil fuel (coal, oil, etc.) and adding its carbon to the atmosphere as carbon dioxide. In 50 years or so this process, says Director Roger Revelle of the Scripps Institution of Oceanography, may have a violent effect on the earth's climate.

Revelle and Hans Eduard Suess (1909–93) warned the following year that the huge quantities of fossil fuel carbon being emitted into the atmosphere would affect the climate. And in 1959, Bert Bolin (1925–2007) reported that the effects of CO_2 emissions on the climate would be extreme.[32] Bolin would go on to become the first chairman of the IPCC.

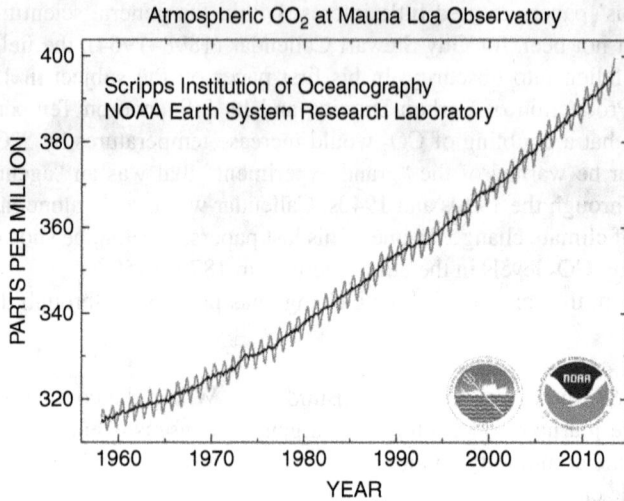

Figure 1.1. Atmospheric CO_2 measurements taken from the Mauna Loa Observatory.

Source: Dr Pieter Tans, NOAA/ESRL (www.esrl.noaa.gov/gmd/ccgg/trends/) and Dr Ralph Keeling, Scripps Institution of Oceanography (scrippsco2.ucsd.edu/); the symbolic 400 ppm mark was surpassed on May 9, 2013 (daily basis) and June 1, 2013 (weekly basis).

Climate change still remained more theory than fact, but that began to change by the end of the decade. A milestone in the science of climate change took place in March 1958 when Charles David Keeling (1928–2005) started collecting data on CO_2 levels in the atmosphere at the Mauna Loa Observatory in Hawaii (part of the National Oceanic and Atmospheric Administration – NOAA). The resultant graphs of atmospheric CO_2 levels are called Keeling Curves (Figure 1.1). The data illustrated the rapid rise of CO_2 emissions, and stunned many observers.

The global scope of climate change invited the participation of the United Nations, with the World Meteorological Organization (WMO) and UNEP leading the early UN effort. A milestone in climate studies had been the First International Conference of Directors of Weather Services in 1873. An outcome of the conference in Vienna was the formation of the International Meteorological Organization. From this group, the World Meteorological Organization[33] was formed in 1950, and the following year, it became an agency of the United Nations. The first Secretary-General was Gustav Swoboda (1893–1956), who was followed by David Arthur Davies (1913–90). During the long tenure of Davies, the WMO formed the Global Atmospheric Research Programme (GARP) in 1967 with the International Council of Scientific Unions (ICSU), and they created the World Weather Watch in 1968 that became a fundamental source of data for climate investigators. And WMO established the World Climate Program.[34] In these early days, Davies warned of CO_2 emissions changing the climate.[35] While scientists had issued such warnings in the past, this proclamation had special significance as Davies represented a United Nations organization. In the last year of his long tenure as Secretary-General, on February 12, 1979, the WMO had organized the first major international conference on climate change, the First World Climate Conference. Aksel C. Wiin-Nielsen (1926–2010)[36] succeeded Davies, and beginning in 1984, the WMO came under the direction of another long-serving Secretary-General, Godwin Olu Patrick Obasi (1933–2007).[37] WMO began working with UNEP on the climate change agenda.

Together UNEP and the WMO would guide the global effort towards Kyoto. In the second half of the 1980s, climate change was thrust into the political spotlight by three initiatives: the Villach Conference in 1985, the *Brundtland Report* in 1987 and the Toronto Conference in 1988. Leading this transformation was the Conference on the Assessment of the Role of Carbon Dioxide and of Other Greenhouse Gases in Climate Variations and Associated Impacts, organized by UNEP, the WMO and the ICSU, which was held at Villach on October 9, 1985.[38]

The Secretary-General of the UN, Javier Pérez de Cuéllar y de la Guerra (1920–), initiated the second action when he requested the former Minister of Environmental Affairs (and briefly Prime Minister) of Norway, Gro Harlem Brundtland (1939–), to organize and lead the World Commission on Environment and Development. In March 1987 the celebrated *Brundtland Report* was issued by the United Nations, which warned of the dangers of CO_2 emissions from fossil fuels:[39]

> There is no way to prove that any of this will happen until it actually occurs. The key question is: How much certainty should governments require before agreeing to take action? If they wait until significant climate change is demonstrated, it may be too late for any countermeasures to be effective against the inertia by then stored in this massive global system …
>
> A safe, environmentally sound, and economically viable energy pathway that will sustain human progress into the distant future is clearly imperative. It is also possible. But it will require new dimensions of political will and institutional cooperation to achieve it.

Brundtland and James William MacNeil (1928–) of Canada, the executive secretary of the Commission, organized a follow-up session[40] – "The Changing Atmosphere: Implications for Global Security" in June 1988 in Toronto, chaired by Howard Ferguson, the Assistant Deputy Minister of the Environment for Canada. Keynote speakers included Brundtland and the Prime Minister of Canada, Brian Mulroney (1939–). A special feature of the Toronto Conference was that it was sponsored by a national government, namely Canada. Recommendations from the conference were:[41]

> Humanity is conducting an unintended, uncontrolled, globally pervasive experiment whose ultimate consequences could be second only to a global nuclear war. The Earth's atmosphere is being changed at an unprecedented rate by … inefficient and wasteful fossil fuel …
>   i. Set energy policies to reduce the emissions of CO_2 and other trace gases, in order to reduce the risks of future global warming.
>   ii. Reduce CO_2 emissions by approximately 20% of 1988 levels by the year 2005 as an initial global goal.

   iii. Set targets for energy efficiency improvements that are directly related to reduction in CO_2 and other greenhouse gases ... Apart from efficiency measures, the desired reduction will require (i) switching to lower CO_2 emitting fuels, (ii) reviewing strategies for the implementation of renewable energy especially advanced biomass conversion technologies; (iii) revisiting the nuclear power option, which lost credibility because of problems related to nuclear safety, radioactive wastes, and nuclear weapons proliferation.

   iv. Initiate the development of a comprehensive global convention as a framework for protocols on the protection of the atmosphere.

Nations were encouraged to set emissions reduction targets. The 20 percent reduction by 2005 was known as the "Toronto Target," which became a euphemism for legally binding CO_2 emissions reduction targets. However, Bert Bolin disagreed with this carbon target as it failed to consider the complexity of such matters.[42] A major outcome of the Toronto Conference was that the public media started to pay attention to climate change.[43]

United Nations Framework Convention on Climate Change

Shortly before the Toronto Conference, the UK had officially called for a framework convention on climate change,[44] and Malta had introduced a draft resolution on October 26, 1988, "The Conservation of Climate as Part of the Common Heritage of Mankind,"[45] which initiated a series of resolutions on climate change in the General Assembly of the United Nations, under the title "Protection of Global Climate for Present and Future Generations of Mankind":

1. *December 6, 1988* (Resolution 43/53)[46] – declared that "climate change" should be a "priority issue," and endorsed the formation of the Intergovernmental Panel on Climate Change (IPCC) to review the science.
2. *December 22, 1989* (Resolution 44/207)[47] – announced a "global framework" to tackle climate change. A critical caveat was added that the "development priorities of the developing nations" had to be taken into account.
3. *December 21, 1990* (Resolution 45/212)[48] – ordered the formation of an "Intergovernmental Negotiating Committee of an effective framework convention on climate change," working with the IPCC.
4. *December 19, 1991* (Resolution 46/169)[49] – urged the formation of "legal instruments" and that the Intergovernmental Negotiating Committee completes their negotiations "as soon as possible."

The road to the United Nations Framework Convention on Climate Change (UNFCCC) was underway. With regard to the UN efforts on climate change, the IPCC became the first official organization formed, which was to prepare the "elements" of the upcoming UNFCCC, and this mandate had been reinforced at the Noordwijk Ministerial Conference on Air Pollution and Climate Change on

November 6, 1989.[50] The WMO and UNEP were planning to take the lead in the actual negotiations[51] and formed the WMO/UNEP Task Force on a Convention on Climate Change, which met for the first time in October 1989.[52] At the following Fourth Session of the IPCC (August 27, 1990), Obasi informed members of the IPCC that in February 1991, national delegates would be invited to the initial meeting for the Framework Convention on Climate Change in Washington, DC (Chantilly).[53] He was still of the opinion that the WMO and UNEP would be coordinating this effort. An organizational meeting to handle the policy negotiations was to be convened in the early fall of 1990.

However, some developing nations were not so happy about the WMO and UNEP handling this process:[54] "Many, essentially developing, countries stressed that the negotiations must be conducted in the forum, manner and with the timing to be decided by the UN General Assembly." In response, another "intergovernmental" body was formed by the United Nations to deal directly with the climate change treaty negotiations which would be only "supported" by WMO and UNEP.[55] Late into the process, on December 21, 1990 (Resolution 45/212), the "Intergovernmental Negotiating Committee of a Framework Convention for Climate Change" (INC) was formed, under the chairmanship of Jean Ripert (1922–2000) of the IPCC.[56] The objectives of the INC were divided into two working groups:[57]

1. *Commitments*: develop principles, objectives and commitments; the Working Group was given the instructions to take "into account that contributions should be equitably differentiated according to countries' responsibilities and their level of development." Working Group I was under the chairmanship of Nobutoshi Akao of Japan and Edmundo de Alba-Alcaraz (1938–) of Mexico.
2. *Mechanisms*: develop the complicated and unprecedented administrative, legal and institutional mechanisms for the implementation of the agreement. Working Group II was under the chairmanship of Robert Van Lierop (1939–) of Vanuatu and Elizabeth Dowdeswell of Canada.

The enormity of the task to prepare an international political accord cannot be emphasized enough, as the members only had fifteen months to work out a deal; they had to be ready for the Earth Summit in Rio. An intrinsic barrier was the economic disparity between developed and developing nations, often called the North–South divide. The Stockholm Declaration in 1972 had outlined the responsibilities between the two groups with regard to environmental issues,[58] which stated that developing countries must be allowed to give priority to their economic development. The ramifications of this decree had far-reaching consequences. At the Second World Climate Conference in Geneva in October 1990 the inseparable link between the environment and the economic development of developing countries became a controversial focal point of this[59] and every INC and COP meeting thereafter. The INC negotiations quickly broke into North and South camps with a clear division between their goals:[60]

The North – developed nations (mainly in the northern hemisphere) do have the greatest responsibilities for climate change, but developing nations should accept some restraints that would increase as their economic position improves.

The South – developing nations (mainly in the southern hemisphere) are victims of the North's emissions; they would have no restraints on their emissions to allow their economies to grow unfettered so they can bring themselves out of poverty. They were largely represented by the G77; this UN group of developing countries had been formed in 1964 (the group now has 130 members). China joined later but at the time was a *de facto* leader of the G77 during the climate change negotiations.

The INC held six (five officially, as the fifth was divided at the last minute into two parts) sessions before the Rio conference:

1. *1st Session* (February 4, 1991) – Chantilly, Virginia (near Washington DC) – the meeting began with remarks by Antoine Blanca, the Director-General for Development and International Economic Co-operation of the UN, who was blunt when describing the importance of the INC; he declared[61] that the United Nations had come into existence because of a "war [WWII] between people. The present meeting of the Intergovernmental Negotiating Committee resulted from war between people and the planet." He stressed the need for unified action: "What was needed at present was to deal not only with the symptoms of climate change, but also with its causes by rethinking the present wasteful style of living. The task before the international community was formidable: rethinking development in the North as well as in the South and ensuring stability and rationality as well as equity and justice between nations, peoples and individuals ... This small planet had started to force mankind to act in concert." Among the options that were debated in this inaugural meeting were:

    *   The "precautionary principle,"[62] whereby scientific uncertainty should not be used as an "excuse" for not taking action on climate change. Touted by the EU, the Americans were reluctant to accept such a policy as they saw climate change legislation as a threat to their economy.
    *   The "polluter pays"[63] based on per capita emissions. Compensation to developing countries should be made for any environmental damages caused by climate change. This was a favorite message from developing countries but, of course, was less attractive to the developed nations.
    *   The general view was to restrict emissions to 1990 levels by 2000. At this session, the controversial principle of "common but differentiated responsibility"[64] was introduced. This phrase had the most profound impact on the negotiations. Over 20 years later, major powers of the world are still debating the implications of these four words.

2. *2nd Session* (June 19, 1991) – Geneva – "common but differentiated responsibility" was discussed as "equitably differentiated according to countries'

responsibilities and their level of development."[65] France supported by Sweden had proposed the establishment of different categories of countries in regard to their "differentiated responsibilities."[66]

3. *3rd Session* (September 9, 1991) – Nairobi – The country category proposal from the second session was put forward again, this time by Italy (and was supported by Sweden and Norway[67]):[68]

   - Least Developed Countries
   - Small Island States
   - Developing Countries – with a substantial industrial sector
   - Developing Countries – oil producing
   - Newly-industrialized Countries
   - Economies-in-Transition (ex. Eastern Bloc countries and Soviet states)
   - OECD (Organisation for Economic Co-operation and Development).

   As nations evolved from one category into another, their commitments would change to match the classification. China immediately objected and would not even consider the Italian proposal, stating that it would relegate "most developing countries to *perpetual poverty*."[69] Kuwait[70] and Saudi Arabia[71] supported the Chinese opposition. The general target of being back to 1990 levels by 2000 for developed countries was put forward, but was again blocked by the US. There was "broad support" for the "precautionary principle," "differentiated responsibility" and the "right to development" but no consensus.

4. *4th Session* (December 9, 2001) – Geneva – An attempt to define "common but differentiated responsibility" was again made, including:[72] "In pursuance of the above global objective, and in accordance with their differentiated responsibilities and taking into account their specific national and development priorities [and objectives and circumstance]/[particularly emission capita, modified as appropriate by other relevant factors] [such as:

   - Emission level per capita,
   - Emission level per unit of GDP,
   - Climatic conditions,
   - Availability of indigenous energy sources, particularly non-carbon sources such as solar, hydropower, biomass, etc.,
   - Industrial structure of the country,
   - Potential for improved energy efficiency and conservation.]"

   The G77 (developing countries) elegantly argued that they should have no emissions targets:[73] "The issues of economic development and equity are vital for the conclusion of a Convention in which the developing nations can participate. Prospects for arriving at a consensus depend on these factors being given the importance they deserve to arrive at a balanced result. The developing countries cannot be expected to accept any formulation that would institutionally help perpetuate, however well intentioned the motives of some parties, the present economic disparity amongst the members of the international community."

A Consolidated Working Document was issued but still heavily "bracketed" to denote areas where objections had been raised.[74] There was agreement that net emissions of developing countries would rise.[75]

Overall, the results of the fourth session were "disappointing," according to Bo Kjellen, the head of the Swedish delegation. The problem was the conflicting positions within the OECD, especially between the EU and the US. As Chandrashekhar Dasgupta of the Indian delegation observed, an agreement could not be reached unless the OECD could "get their act together."[76]

5a.   *5th Session* (Part I, February 18, 1992) – New York – "Common but differentiated responsibility" was still under intense debate.[77] A "revised text under negotiation" was issued, which contained 800 brackets![78] Drafts became illegible.[79] So much work remained, mainly caused by the tactics of the Americans, that a second session was needed.[80]

5b.   *5th Session* (Part II – April 9, 1992)[81] – New York – A skillfully drafted document had been prepared by the Chairman, Jean Ripert,[82] which the delegates accepted. There was no clarification of the meaning of "common, but differentiated responsibilities."

The world's leading political negotiators had hashed out an agreement, clause by clause, sentence by sentence, and even word by word (not to mention, bracket by bracket), but a major issue could not be resolved – the definition of "common but differentiated responsibility" – which would continue to burden all future climate change negotiations. But time had run out and a weak agreement was better than no agreement at all; a treaty was ready to be signed at Rio.

The environmental agreement was officially named the United Nations Framework Convention on Climate Change (UNFCCC).[83] The grand climate manifesto began:[84]

The Parties to this Convention,

Acknowledging that change in the Earth's climate and its adverse effects are a common concern of humankind,

Concerned that human activities have been substantially increasing the atmospheric concentrations of greenhouse gases, that these increases enhance the natural greenhouse effect, and that this will result on average in an additional warming of the Earth's surface and atmosphere and may adversely affect natural ecosystems and humankind.

Noting that the largest share of historical and current global emissions of greenhouse gases has originated in developed countries, that per capita emissions in developing countries are still relatively low and that the share of global emissions originating in developing countries will grow to meet their social and development needs.

The objective and first principle were:[85]

[S]tabilization of greenhouse gas concentrations in the atmosphere at a level that would prevent dangerous anthropogenic interference with the climate system. Such a level should be achieved within a time frame sufficient to allow ecosystems to adapt naturally to climate change ...

The Parties should protect the climate system for the benefit of present and future generations of humankind, on the basis of equity[86] and in accordance with their *common but differentiated responsibilities* [emphasis added] and respective capabilities. Accordingly, the developed country Parties should take the lead in combating climate change and the adverse effects thereof.

As the delegates arrived at the "Earth Summit"[87] in Rio de Janeiro on June 3, 1992, there was a festive atmosphere over the epic document. And after the delegates approved this green blueprint came the usual photo-ops and press interviews. Despite all the hoopla, the nations of the world did not agree to anything meaningful in Rio. There was agreement to reduce emissions to 1990 levels by 2000 in principle, but these were aims with no legally binding implications, which meant that nothing would be done. Some nations, especially the US, hoped to keep it that way. The only real accomplishment was that the parties agreed to keep talking and talking and ... over 20 years later, they are still talking while emissions are still rising.

Kyoto Protocol

The INC, COP, UNFCCC and IPCC are not household names (more properly acronyms), but Kyoto is synonymous with climate change negotiations to many. The name originated from the site of COP 3 which took place at the end of 1997. After the framework convention was agreed to at Rio in June 1992, the real work began negotiating legally binding protocols; this task would be the job of the Conference of the Parties (COP) meetings.[88] However, many procedural issues remained that should have already been handled in the previous meetings of the INC, including the pesky definition of "common but differentiated responsibility." In an unusual move, the INC kept on meeting after Rio was signed, as they did their best to deal with the unfinished business of various administrative and policy issues, meeting six more times (their eleventh session was held February 6, 1995).

Even with the extra sessions of the INC, some major issues had still not been ironed out. At INC 10, delegates had decided that the decision on the protocol would be delayed until COP 3 at Kyoto. Consequently, the first COP meetings in Berlin[89] and then Geneva would have to settle them. In April 1995, the inaugural COP was held in Berlin; where once East met West in the Cold War, now North met South in the Hot War over global warming. The most important outcome of COP 1 was a bizarre resolution of the problematic "common but differentiated responsibilities" through the exemption clause [2 (b)]:[90] "Not introduce any new commitments for Parties not included in Annex I."[91]

By this simple statement, the developing countries were excluded from any targets. Even though the actual phrase "common but differentiated responsibilities" was never clarified, the practical result was that it no longer mattered. The Berlin Mandate gave the developing countries what they were seeking: no targets at all. At COP 2 in Geneva in July 1996, the declaration of the Berlin Mandate was reaffirmed.[92]

On December 11, 1997, the Kyoto Protocol was adopted,[93] with the Berlin Mandate firmly embedded in it.[94] Agreement was reached to reduce greenhouse gases by 5.2 percent from 1990 levels for the developed nations overall. The Kyoto targets varied regionally ("differentiated") but the bulk of the global reduction was assigned to the EU (with an 8 percent decline) and the US (with a 7 percent decline) in GHG emissions, which were to be achieved over the average of the years 2008–12.[95] This period had been promoted by the American delegation, as the Europeans had been looking for an earlier time frame.[96]

However, the Americans would never ratify the agreement. Essentially, the Kyoto negotiations brought in the 150 or so "free-rider" nations at the expense of one nation in particular. Unfortunately, this one nation, the US, was the largest CO_2 emitter in the world at the time; and the second largest emitter, China (now the largest), was already exempt as a developing nation. In the process of winning an agreement, under great international pressure, and through many years of contentious and intense negotiations, the Kyoto Protocol, with all its compromises, was no more than an exhibition game. At least, however, some of the teams were playing.

Relationship of the INC & COP with the IPCC

When the UN turned to fill the membership of the INC, many of the members had come from the IPCC, including the chairman, Jean Ripert, and Working Group Chair, Elizabeth Dowdeswell. The IPCC was to serve the new group's needs.[97] While Bolin, the Chairman of the IPCC, agreed to aid the INC, he appeared cautious about the new role the IPCC was having with the treaty negotiations. He emphasized that the IPCC was still an "independent body."[98]

There was friction between the two intergovernmental cousins. The INC needed help from the IPCC quickly, as they had deadlines for upcoming international negotiations that would depend upon the Assessment Reports from the IPCC. But the IPCC had been designed for detail and accuracy, not speed; the writing teams had to deal with a huge amount of information which required time to properly review and to assess in the manner demanded by the government members of the IPCC. Somehow the two schedules had to come together.

Tolba set the stage by putting pressure on the members to act quickly at the First Session of the IPCC, for the First Assessment Report (FAR) was needed at the upcoming Second World Climate Conference[99] on October 29, 1990 (the original schedule had called for the Conference on June 25 but this was delayed to give the IPCC more time to complete its first assessment). And then the Supplement Report was required for the Earth Summit in Rio. At the Sixth Session of the IPCC (October 1991), Bolin reminded members that time was running out as the Earth Summit was quickly approaching.[100]

Despite the tight deadlines, the IPCC met its early commitments and was a prime motivator for the UNFCCC agreement at the Earth Summit in Rio in 1992:

- Michael Zammit Cutajar, Executive Secretary of the INC, praised the IPCC for their contributions.[101]
- Jean Ripert, who led the INC and had been a prominent member of the IPCC, commented on how the education of negotiators by the IPCC was key to the agreement.[102]
- Yvo de Boer, Executive Secretary of the UNFCCC, later stated that the work of the IPCC was, in many ways, responsible for the birth of the UNFCCC.[103]

Article 21 of the UNFCCC declared that the INC would continue to work with the IPCC,[104] and the Assessment Reports of the IPCC were to be based around Article 2,[105] which stated:[106]

> The ultimate objective of this Convention and any related legal instruments that the Conference of the Parties may adopt is to achieve, in accordance with the relevant provisions of the Convention, stabilization of greenhouse gas concentrations in the atmosphere at a level that would prevent dangerous anthropogenic interference with the climate system. Such a level should be achieved within a time frame sufficient to allow ecosystems to adapt naturally to climate change, to ensure that food production is not threatened and to enable economic development to proceed in a sustainable manner.

In June 1992 at the 44th Session of the Executive Council of the WMO, the IPCC was formally requested to provide support of the newly negotiated UNFCCC,[107] which the IPCC accepted.[108]

After Rio, the international negotiations on climate change entered a new phase, and more pressure from the INC was directed at the IPCC. Bolin had sent a note summarizing the action plans of the IPCC to Ripert on March 1, 1993. However, Ripert was no longer the chairman of the INC, as he had just been replaced by the career-diplomat Raúl Estrada-Oyuela. Estrada-Oyuela responded[109] to Bolin on March 18, and addressed the IPCC at the Ninth Session (June 1993), where he gave a long list of requirements from the IPCC.[110] Bolin confirmed that it was of the "utmost importance" to respond to the needs of the INC,[111] but that the IPCC would "maintain its flexibility to exercise judgment of what was essential."[112]

To deal with this issue of timing and other issues, the IPCC and INC formed the Joint Working Party (JWP), later becoming the Joint Working Group under the COP, which was composed of senior members of both organizations. The JWP first met in November 1993.[113] Initial meetings, though, failed to defuse the tension. Estrada-Oyuela became increasingly irritated with the lack of responsiveness by the IPCC, though Bolin downplayed the disagreement.[114] Estrada-Oyuela even publicly chastised the IPCC at the Royal Geographical Society on May 31, 1994.[115] Here he made his famous witty statement that the IPCC was suffering from the "Dr Frankenstein syndrome." The story was picked up by the popular science magazine, *New Scientist*, under the banner "Frankenstein Syndrome Hits Climate Treaty"; the article began:[116]

> Politicians and scientists are at loggerheads over the Climate Change Convention, the most important outcome of the Earth Summit at Rio in 1992. The head negotiator for implementing the convention, aimed at combating the greenhouse effect, last week accused scientists of 'suffering from a Dr Frankenstein syndrome'.
>
> The work of scientists created the convention. But 'now the convention is alive and walking, and deciding things, the scientists have reacted against its demands', claimed Raul Estrada-Oyuela.

The first IPCC Plenary Session (10th) after the Frankenstein syndrome comment took place a few months later. Estrada-Oyuela did not attend but the INC representative Michael Zammit Cutajar reminded the members of outstanding issues and mentioned the "timely availability of inputs" for the INC.[117] Near the end of his address, Cutajar stated that the two bodies must continue to work closely together for the IPCC to remain "relevant." Bolin was not intimidated, and he was not going to allow the INC to run the IPCC; he stated the IPCC had to decide what the relationship was going to be with the INC and that a decision would be made at the following session. This was the last IPCC Session before the first meeting of the Conference of the Parties in Berlin (COP 1).

The IPCC continued to support the Conference of the Parties (COP);[118] but the IPCC became more distant from the actual negotiations as it no longer dealt directly with the senior members of the COP; contact between the two organizations now took place through the Subsidiary Body for Scientific and Technological Advice (SBSTA), which was the technical arm of the UNFCCC.[119]

The transition from the INC to the COP and SBSTA did nothing to lower the tensions between the scientists and the negotiators. After COP 1, in the minutes of the Eleventh Session (December 1995) of the IPCC, the time constraints of the COP are highlighted.[120] The notes imply that the IPCC resented the added pressure but that they would adjust their procedures to be more amenable to the COP. However, the peer-review process that lay at the heart of the credibility of their reports prevented the IPCC from reacting in too hasty a fashion to short-term requests by the COP.[121]

At the 22nd Session of the IPCC there was again debate over timing. The UK warned that no Synthesis Report for COP 13 (Bali, December 2007) could make the IPCC "irrelevant,"[122] a threat that the IPCC had heard before.

At the same session, Germany told members of the IPCC that the COP was the organization's "main customer," and at the 35th Session (June 2012) an interesting debate was held on who was the customer of the IPCC:[123]

Switzerland, opposed by the US and Canada, also suggested mentioning 'in particular' the UNFCCC as the main user of IPCC products. The US stressed that IPCC documents are primarily aimed at governments. Canada indicated that the IPCC's relevance for the UNFCCC will be determined by the quality and timeliness of its work rather than by the language of the Strategy.

Intergovernmental Panel on Climate Change (IPCC)

The formation of the UNFCCC and the negotiations of the INC and COP depended on the studies of the IPCC, whose reports are the result of the collaboration of more than 1,000 scientists, working on a voluntary basis, who scan the scientific literature on climate change and present summaries in layman's terms.[124] The evolution and structure of this unique knowledge-philanthropic group, and the factors that have affected its authority, will be examined.

The seed of the IPCC was planted at Villach.[125] After the conference, Tolba urged George Shultz (1920–), the US Secretary of State under Reagan, to adopt policies to combat climate change. The US government adopted a stalling tactic and proposed an "intergovernmental mechanism" to study the science behind climate change.[126] The IPCC was born out of this compromise between the forces supporting climate change legislation and those hostile to it.

At the Tenth WMO Congress in May 1987 the path towards the IPCC was officially underway. Resolution 9 – Global Climate Change – was passed on June 18,[127] "to arrange for *appropriate* mechanisms to undertake further development of scientific and other aspects of greenhouse gases." And the following week, the WMO Executive Council called upon the Secretary-General of the WMO, "in coordination with the Executive Director of UNEP to establish an *intergovernmental* mechanism to carry out internationally coordinated scientific assessments of the magnitude, impact and potential timing of climate change."[128] The design of the new organization took place over the following months, under the influence of the WMO, UNEP and various US government agencies.[129] On March 25, 1988, Obasi formally announced the formation of the Intergovernmental Panel on Climate Change (IPCC) to its members.[130] And, in June, at the 40th Session of the Executive Council of the WMO, Resolution 4 announced the official creation of the Intergovernmental Panel on Climate Change:[131]

> That the activities of the Intergovernmental Panel on Climate Change (IPCC) should be aimed at (i) assessing the scientific information that is related to the various components of the climate change issue such as emissions of major greenhouse gases and modification of the Earth's radiation balance resulting therefrom, and that needed to enable the environmental and socio-economic consequences of climate change to be evaluated; and (ii) formulating realistic response strategies for the management of the climate change issue.

News traveled fast as the fledgling organization was recognized in the same month at the Toronto Conference.[132]

United Nations recognition of the IPCC came shortly afterwards. Among the various policy initiatives of UN Resolution 43/53, in late 1988, was support for the IPCC: the UN General Assembly,[133]

> Endorses the action of the World Meteorological Organization and the United Nations Environmental Programme in jointly establishing an Intergovernmental Panel on Climate Change to provide internationally coordinated scientific assessments of the magnitude, timing and potential environmental and socio-economic impact of climate change and realistic response strategies ...
>
> Requests the Secretary-General of the World Meteorological Organization and the Executive Director of the United Nations Environmental Programme, utilizing the Intergovernmental Panel on Climate Change, immediately to initiate action leading, as soon as possible, to a comprehensive review and recommendations with respect to:
> a)   The state of knowledge of the science of climate and climatic change;
> b)   Programmes and studies on the social and economic impact of climate change, including global warming;
> c)   Possible response strategies to delay, limit or mitigate the impact of adverse climate change;
> d)   The identification and possible strengthening of relevant existing international legal instruments having a bearing on climate;
> e)   Elements for inclusion in a possible future international convention on climate.

A formal Memorandum of Understanding (MoU) on the IPCC had been signed by Jim Bruce of the WMO on April 14 and Anthony Brough of UNEP on May 8, 1989, where the two groups agreed to financial arrangements for the IPCC.[134] Officially, the IPCC was an independent body under the auspices of the WMO and UNEP:[135]

> From an institutional perspective, IPCC is a joint panel of UNEP and WMO, as stipulated in decision SS.II/3 B of 3 August 1990 of the UNEP Governing Council and Resolution 4 (EC-XL) of the WMO Executive Council. The IPCC is integral part of the institutional arrangements of these two organizations.

A later review found that:[136]

> The IPCC has reporting responsibilities to four United Nations bodies: UNEP, WMO, the United Nations Framework Convention on Climate Change (UNFCCC), and the UN General Assembly. Legally, the IPCC is an intergovernmental joint subsidiary panel of WMO and UNEP, but it has operated in practice as an independent organization. Perhaps as a consequence, although strongly supportive of the IPCC, WMO and UNEP officials appear to exert modest oversight over the organization.

The WMO and UNEP, then, had a very high-level role to play with the IPCC, but not much more. The two UN bodies had been instrumental in the creation of the IPCC and were often referred to as the parent or sponsoring organizations. The IPCC clearly needed their official endorsement to continue to operate,[137] but they had little control or direct influence over the IPCC.[138] Numerous resolutions by the WMO and UNEP were served upon the IPCC, but the IPCC often decided not to act upon them.[139] For example, even when requested by the WMO Congress in 2003 to reconsider the important Terms of Reference, the IPCC stated that it was not necessary. In a candid statement, Bolin felt that the two parent organizations viewed the IPCC as a threat because of its independence.[140]

Reports

The IPCC has access to a volunteer army but has no direct control over them; it is an umbrella organization, a networking group of the leading climate scholars in the world, which translates the scientific literature into relevant conclusions for policymakers. The IPCC does not carry out scientific experiments, measure CO_2 levels in the atmosphere, fund research or produce any new knowledge on climate change. Their mandate is to prepare assessments of the literature. The output of the IPCC includes:[141]

- Methodology Reports – guidelines on the determination of greenhouse gas inventories;
- Special Reports – prepared upon request of the WMO/UNEP, INC or COP. The *Supplement Report of 1992* and *Climate Change 1994* fall into this category; the latest being the *Special Report on Managing the Risks of Extreme Events and Disasters to Advance Climate Change Adaptation* (2012), *Special Report on Renewable Energy Sources and Climate Change Mitigation* (2011) and *Carbon Capture and Storage* (2005);
- Technical Papers – fast-tracked communication products for the COP, based on information from already-issued IPCC Reports.[142] The Bureau had the final say on their contents, and the documents would not be officially "accepted" or "approved" by the IPCC. The first of these papers to be issued was *Technologies, Policies, and Measures for Mitigating Climate Change*. Other early papers were prepared for COP 3 at Kyoto and presented by Bolin and Watson:[143]
  1. *Simple Climate Models*;
  2. *Stabilization of Greenhouse Gas Concentrations in the Atmosphere*;
  3. *Environmental Implications of Emission Limitations*.

The *raison d'être*, though, of the IPCC, was the Assessment Report. More than 1,000 of the world's leading climate scholars review more than 10,000 references every six years or so. A shorthand notation was adopted for the naming of the Assessment Reports:

> First Assessment Report – FAR
> Second Assessment Report – SAR
> Third Assessment Report – TAR

The creators of the original nomenclature likely never foresaw that a Fourth Assessment Report would be necessary. Since the acronym labeling would no longer work, a new naming system had to be created. At the 27th Session of the Bureau (November 2007), various options were suggested: FoAR, 4AR and AR4.[144] The latter, "AR4," was adopted as the new naming format.

An important aspect of the output of the IPCC is illustrated in the title of these documents; they are not called reviews for a reason.[145] The IPCC brings together the current state of knowledge of climate change science but placed in the context of the practical needs of the policymaker; critical issues are identified and the confidence of conclusions estimated. They are also not supposed to favor any particular policy to combat climate change. The language is carefully controlled, especially by policymakers, not to inflame public opinion or to speculate beyond what the literature directly illustrates. The bulk of the Assessment Report is the basic scientific review, whose main audience is specialists in the field, and academics.

The section with broader appeal, especially to the policymakers who were the intended customers of the reports, is the Synthesis Report. The Synthesis Report was defined as:[146]

> **"Synthesis Reports"** synthesise and integrate materials contained within the Assessment Reports and Special Reports and are written in a non-technical style suitable for policymakers and address a broad-range of policy relevant but policy-neutral questions. They are composed of two sections as follows: (a) a Summary for Policymakers and (b) a longer report.

The Summary for Policymakers (SPM) had been introduced at the Second Session of the IPCC,[147] and John Houghton described the SPM as being for "those without a strong science background."[148] The premise of the Synthesis Report was that it was relevant to Article 2 of the UNFCCC, but the specific format to carry out this task varied.[149] It first appeared as part of the Second Assessment Report (SAR)[150] and was published as a separate document in the Third Assessment Report (TAR).[151] COP 3 (December 1997) requested a change in the format; the report was to focus on policy-relevant questions. Nine questions submitted by governments were chosen by the IPCC for the TAR Synthesis Report.[152] For the Fourth Assessment Report (AR4), it was decided to use a similar format as the TAR, but with simpler questions.[153]

There are three sets of Assessment Reports, one from each Working Group (WG), assisted by the Technical Support Units (TSUs – this staff of a few people aid with administrative and communications duties for the writing teams). The original UN mandate, Resolution 43/53, had listed five areas of responsibility for the IPCC, but the latter three were related to mitigation and responses. At the First Session of the IPCC, Tolba[154] suggested the establishment of three Working Groups:[155]

Working Group I (WG I) on Scientific Assessment
Working Group II (WG II) on Potential Impacts
Working Group III (WG III) on the Formulation of Response Strategies.

This model of groups reporting on science, impacts and mitigation (response) would, with minor changes and a few exceptions (especially the SAR), hold throughout the years of the IPCC. WG I and WG II dealt with climate change with regard to risk assessment, while WG III dealt with risk management. By its nature, the mandate of WG III was more judgmental and was more prone to "politicization."[156] The UN mandate had also added the responsibility of reviewing the less scientific socio-economic consequences of the impact and mitigation of climate change.

The preparation of the Assessment Reports by the WGs involves a complex structure,[157] with a number of roles and responsibilities:

- *Working Group Chairs and Vice-Chairs* – coordinate activities and appoint the Coordinating Lead Authors (can also be chosen by other members of the IPCC Bureau). They are also responsible for preparation of the Summary for Policymakers (SPM).[158] They are members of the IPCC.
- *Coordinating Lead Authors* (CLAs) – are Lead Authors, who have the added responsibility for putting together a specific chapter (two CLAs per chapter). They act as editors, being responsible for submitting the draft chapter to the Working Group Chairs. The CLAs are often members of the IPCC and are chosen by the IPCC Bureau.
- *Lead Authors* (LAs) – are responsible for producing a section of a chapter. Along with the Review Editors, they are also responsible for taking into account the Expert Reviewers' and Government Reviewers' comments. Lead Authors may also solicit Contributing Authors. They can be members of the IPCC, but not necessarily so. The nominations come from governments, individual scientists and international scientific and industrial organizations;[159] among the thousands of nominations, the IPCC Bureau makes the final choice.
- *Contributing Authors* (CAs) – are specialists who provide information from "peer reviewed and internationally available literature" on specific sections of the chapters to the Lead Authors. They are chosen by Lead Authors (and CLAs) with general approval by the Working Group Bureau.
- *Review Editors* (REs) – ensure that the review process and comments on the drafts of the Assessment Reports are handled according to the rules and procedures of the IPCC; they coordinate activities between the Expert Reviewers and the Lead Authors. There are one or two REs per chapter, who are chosen by the IPCC Bureau.
- *Expert Reviewers* (ERs) – are nominated from the above positions, governments and other organizations; general criteria for their selection are listed in the procedures for preparation of the Assessment Reports.[160] Often, they have been nominated in the list of Lead Authors (but not chosen), but could also

be Lead Authors from other chapters. Each section is submitted to Expert Reviewers by the Lead Authors.[161]

- *Government Reviewers* – upon completion of the second draft of the Assessment Report, it is sent out to member countries for "further scientific peer review"[162] through the Government Focal Points.[163] Reviewers are assigned by the government and are often bureaucrats in government departments and ministries.

The IPCC invites the member states (nominations are also sought from intergovernmental and non-governmental organizations) to submit their recommendations for national experts to serve on the writing teams for the reports, under the following conditions:[164]

[T]he IPCC requires that the nominee(s) have appropriate expertise. All nominations should indicate the working group(s), chapter(s) and author role(s) for which the candidate is being nominated and be accompanied by a summarized curriculum vitae, including a specification of the nominees expertise and a list of relevant publications.

In addition, member states were informed in advance on experts invited from their countries.[165]

The writing teams participate in various stages of the Assessment Report process:[166]

1. Structure & Focus of the Assessment Report
    i. Scoping Meeting for Assessment Report (Terms of Reference) – where the structure, work programs and timetables for the Working Groups (WGs) are determined (held about five years before Report is to be adopted by the IPCC); important scientific issues are identified and general outlines of chapters are discussed;
    ii. Scoping Meeting for Synthesis Report including cross-cutting areas;[167]
    iii. Plenary Session – approval of the work program of the Assessment Report is given at the Working Group Plenary Session and then the IPCC Plenary Session.
2. Writing Team Selection (CLA, LA, CA, RE and ER)
    i. Government, participating organizations, IPCC Bureau and WG Bureau prepare nominations for the team (compilation of acceptable contributors who are experts and can contribute to each section of the report);
    ii. WG Bureaus select the CLAs and LAs; governments and participating organizations are informed of CLA and LA choices;
    iii. WG Plenary Session approval of the CLAs and LAs;
    iv. CLAs choose writing teams for their chapter.
3. Drafting of the Assessment Report
    i. Workshops and expert meetings (spread over about one year) are held;
    ii. LA meetings are held (first one about a year after the first scoping meeting);

Table 1.1. The metrics of the assessment reports of WG I.[313]

	FAR	*AR4*
Pages	365	987
CAs		>400
LAs		170
CAs+LAs	160	>570
CAs+LAs: Countries	25	40
ERs	200	600
Final Plenary: Countries	35	113

Source: Renate, 2010, slide 17.

    iii. LA draft-1 issued for review by ER (about a year after the first LA meeting);
    iv. LA draft-2 issued for review by ER and governments (about a year after LA draft-1);
    v. LA draft-3 (final) issued by CLA and LA for review by governments.
    vi. Bureau prepares Synthesis Report draft.
4. Approval
    i. WG Plenary Session approves the report (about a year after the LA draft-2);
    ii. IPCC Plenary Session accepts WG reports;
    iii. IPCC Plenary Session reviews and approves Synthesis Report.

Quality control for the Assessment Report process was peer review,[168] which was integral to the "technical-scientific integrity" of the IPCC.[169] But this was no ordinary academic process, as it was taken to a new level by the IPCC.[170] Peer review has existed in scientific circles for a long time, especially in journals. Usually the peers are few. The Assessment Report is reviewed by thousands. This is not peer review; it is mob review. Although the quantity of comments increases, the overall quality of the review process does not necessarily improve. Such a behemoth process places extra pressure on the Lead Authors to handle the commentary overload.

Already unprecedented in scale when the FAR had been prepared, the task had only increased with each new Assessment Report, reaching new heights with the AR4 (Table 1.1). Overall, the AR4 encompassed more than 450 LAs, 800 CAs and 2,500 expert reviewers. The volume of literature increased greatly over the Assessment Reports:[171]

SAR – 4,000+; TAR – 10,000+; AR4 – 20,000+

The Chairman of the IPCC estimated that the number of references in the AR5 will be greater than 60,000.[172]

Plenary sessions

November 9, 1988 – Geneva – 10:15 AM: history was made as the First (Plenary) Session of the IPCC was opened by Jim Bruce of Canada, the Acting Deputy-Secretary General of WMO,[173] who had been the chair of the Villach Conference in 1985. Obasi and Tolba then welcomed the 90 representatives from 30 countries, plus another 40 representatives of various UN organizations and NGOs.[174] No one at that historic meeting would have foreseen that Plenary Sessions would still be around 25 years later!

At the Plenary Sessions, the government-appointed members would discuss major issues, such as organizational changes, future action plans, election of the Bureau and, most importantly, sanction of the reports:[175]

- "acceptance" of IPCC reports at a session of the Working Group or Panel signifies that the material has not been subject to line by line discussion and agreement, but nevertheless presents a comprehensive, objective and balanced view of the subject matter;
- "adoption" of IPCC reports is a process of endorsement section-by-section used for the longer report of the Synthesis Report;
- "approval" of IPCC Summaries for Policymakers (SPM) signifies that the material has been subjected to detailed, line by line discussion and agreement.

The strict designations imply that they signify something of importance. However, the practical aspect was that they were only an internal matter, as the designations were bureaucratic in nature.

Besides the main Plenary Session of the IPCC overall, each of the three Working Groups holds its own Plenary Sessions. The highlights of the Working Group Plenary Sessions were the sanction of the Assessment Reports and the line-by-line "approval" of the SPM.[176] Since all members had the right to participate in the Working Groups, their "approval" was taken as approval by the IPCC. The Assessment Reports only receive official "acceptance" at the IPCC Plenary Session.

These challenging meetings are where some of the more intense confrontations take place between scientists and politicians:[177]

> The Working Group plenary sessions become the occasions when the scientists engaged by the IPCC are confronted with political views and even attempts to clothe special interests in scientific terms. It is essential that the line between factual information and political judgement is upheld at that time ... Political negotiations should build on factual presentations of available knowledge.

A member of the IPCC provided an in-depth description of what happens at the Plenary Session (for the approval of the FAR of WG I):[178]

> [H]aving started in a very organized fashion with songs about the future from children's choirs ... the meeting came close to a breakdown. It finished at four o'clock in the morning, one day late, with most of the delegates having

abandoned their chairs in the conference hall to gather on the front podium and shout at each other.

The intense political drama continued at the full Plenary Session:[179]

At the final IPCC meeting in Sweden, key recommendations of the summary report, were intensely debated, and on the last night, the meeting came close to failure. Individual countries and groups of countries began to position themselves for the upcoming negotiations. Certain words began to take on enormous importance. For example, in an effort to protect their interests, OPEC countries for the first time argued strongly for the wording 'safe' anytime it was used in reference to nuclear power or energy efficiency. Russia, especially unhappy about the conclusions of the scientific report, argued for inclusion of Budyko's paleoclimate scenarios. India, speaking for the G-77, insisted on language that put the problem of climate change squarely on the backs of industrial countries. The final hours of the IPCC were not a pretty picture.

The official description of the approvals imply a business-like atmosphere; for example, for the TAR of WG I at the Group Plenary Session held in Shanghai in early 2001:[180]

All the comments received were carefully analysed and assimilated into a revised document for consideration at the session of Working Group I held in Shanghai, January 17 to 20, 2001. There the Summary for Policymakers was approved in detail and the underlying report accepted.

The actual meeting, though, could best be described as organized chaos:[181]

However, in Shanghai, it appeared that there were attempts to blunt and perhaps obfuscate the messages in the report, most notably by Saudi Arabia. This led to very protracted debates over wording on even bland and what should have been uncontroversial text. For instance, the opening introductory paragraph of five lines took more than an hour and a half to work out, and then with a small group spun off to recommend revised wording. The chairman, Sir John Houghton of the United Kingdom, and those who helped run the meeting showed fortitude in their unflagging commitment to continue without taking a break. On the second day, following a one-and-a-half-hour break to quickly consume the evening meal, the session ran until 10:45 p.m. The days began for most of the participants at 8:00 a.m. with sidebar meetings. Adjournment on the third day was at 11:30 p.m., and the SPM was finally approved at 12:45 a.m. on the final day. The meeting closed just after 1:00 a.m. In many ways the meeting became one of endurance.

Bureau

A key body between the scientists and the politicians is the Bureau. The Bureau assists the Chairman on issues regarding the procedures for the Plenary Sessions, the work of the IPCC and financing from the IPCC Trust Fund.[182] Membership of the Bureau changed over the years but the core group included:

1.   Chairman of the IPCC;
2.   Vice-Chairmen of the IPCC;
3.   Chairmen of the Working Groups;
4.   Vice-Chairmen of the Working Groups.

The number of Bureau members was originally set at 15[183] but later doubled in size.

The most important position of the IPCC and the Bureau was the Chairman, who proposes the agenda and presides over the Bureau and the Plenary Sessions; especially at the latter meetings, the Chair plays a pivotal role in negotiating settlements between political agendas and scientific integrity during the approval of major documents such as the SPM of the Synthesis Report. Another responsibility of the Chairman is to report back to the WMO and UNEP, and the INC and COP on the activities of the IPCC.[184] More importantly, the Chairman is the public face of the IPCC.

Bert Bolin, the first Chairman, ably guided the IPCC through its early years, and his leadership contributed to its initial success. His challenge was immense as such an organization as the IPCC had not existed before. Bolin would oversee the IPCC for 13 Plenary Sessions. For his contribution, he was unanimously elected as Chairman Emeritus.

The year 2007 was a time of reflection for the IPCC. They lost two of their founders: Godwin Obasi of WMO on March 3, and Bolin on December 30. A few weeks before the death of Bolin, the IPCC had been awarded the 2007 Nobel Peace Prize on December 10. At the 28th Session of the IPCC in April 2008, a tribute was paid to the Chairman Emeritus.[185]

Bolin had carefully sculptured the IPCC to withstand political and bureaucratic meddling in its scientific pursuits. He was passionate about any outside interference. As a result, the inner workings of the IPCC were not widely publicized. Among the puzzling aspects of the IPCC were the elections of the Bureau, especially the Chairman. Members elected (by acclamation) to the Bureau for the First Session were Bert Bolin of Sweden as Chairman, Abdul Bar Bin Al-Gain of Saudi Arabia as Vice-Chairman and Kolawole R. Rufai of Nigeria as Rapporteur (the latter was replaced by Joseph A. Adejokun of Nigeria).[186] With Bolin twice, and then Watson (his successor), the Chairman had just been affirmed at the IPCC Plenary Session. There was no voting as there were no other official candidates.

At the Twelfth Session (September 1996), Robert Tony Watson (1948–) of the United States became Chairman, succeeding Bolin.[187] Again no election had been held. India had stated it would be putting forth a candidate but one was never made to the Nominations Committee.[188] The unanimous decision to elect Watson was due to Bolin's support.[189] His nomination had been officially proposed by Rosina Bierbaum (1952–) of the US:[190]

> Dr. Watson sees the need to address the science, environmental impacts, and socioeconomic aspects of issues in an integrated fashion. More importantly, he sees the balances among the roles of developing and developed nations, as well as among governments, industry and environmental organizations …
>
> And that is exactly what IPCC needs – someone who can represent all of us.

Watson encouraged the IPCC to be more involved in the "science–policy divide." While he supported the IPCC to be policy relevant but not "policy pre-scriptive," he also recognized that the "line that divides the two is often quite narrow."[191] He was, at first, encouraging members to push the envelope. However, in the final draft of his vision for the IPCC, the message was more cautious:[192]

> The IPCC decides that its work must continue to maintain its high scientific and technical standards, independence, transparency and geographic balance, to ensure a balanced reporting of viewpoints and to be policy relevant but not policy prescriptive or policy driven.

Watson had been with the US Office of Science and Technology Policy of the Office of the President, a Lead Author of the FAR of WG I and the Co-Chair of WG II for the SAR. In 2012, Watson was knighted.

Election results appeared for the first time in the published minutes of the Twelfth Session (September 1996). For the Vice-Chairman of the IPCC, repre-senting Asia, Rajendra Pachauri had been elected; he had received 71 votes, while his opponent, Nizar Tawfik of Saudi Arabia, received 35.[193] Pachauri was an elec-tion trend setter within the IPCC. For, during the next election, Pachauri became the first Chairman to actually be elected at a Plenary Session; this was the only time that a Chairman had been elected by a true democratic vote in the 25-year history of the IPCC.

The election of Pachauri as Chairman, however, had been controversial. At the Nineteenth Session in 2002, the incumbent Chairman lost (the other members of the Bureau had been elected by acclamation);[194] the results were Pachauri of India (76), Robert Watson of the United States (49) and José Goldemberg (1928–) of Brazil (7). Many members noted during the election of Pachauri that the proce-dures for elections were not clear.[195] Two weeks after the vote, an article appeared in *Science* by Bolin. He attributed Pachauri's election to the desire of developing countries to have one of their own running the IPCC. Bolin, who had returned to head the Nominating Committee for the election, then warned:[196] "Participation by scientists and experts in developed countries, where much of the relevant basic research and technical development is carried out, must be secured." He was con-cerned by the turn of events:[197] "I was disappointed about the outcome of the vote and felt that the intensity of the IPCC activities might decline."

Pachauri, the new Chairman of the IPCC, was trained as an engineer and worked as an economist; he was the Director-General of The Energy and Resources Institute (TERI) in New Delhi. TERI is a non-profit research organization, which focuses on energy, the environment and sustainable development.[198] Before his election as Chairman, Pachauri had been the IPCC Vice-Chair and Chairman of the *Ad Hoc* Group on the IPCC Communication Strategy.

Pre-election negotiations had generally settled who was going to get the position of Chairman.[199] True elections were rare. There was a behind-the-scenes group that controlled the choice of candidates. A Nominations Committee decided on who should appear "on the slate of candidates for election."[200] The group is only vaguely described in the public documents of the IPCC, and many members of the IPCC did not understand how it worked. China, for example, questioned the power of this committee; in response, the Nominations Committee was stated not to have the right to exclude anyone from the list, but nowhere in the actual rules was this mentioned at the time.[201] The issue was only later clarified at the 24th Session of the IPCC (September 2005) in Rule 21, which opened the election slate to any nominee that has the support of a member.[202] However, in Annex B, the Nominations Committee is clearly given the mandate to choose who appears on the final nominee list.[203] The US and other nations also raised the issue of its power.[204] Disturbingly, delegates were still unclear of the function of the Nominating Committee as late as 2009.[205] Rules 21 and 22 were under active review in late 2011,[206] and new election rules were issued in June 2012; Rule 22 still stated that the "Nominations Committee shall prepare and submit through its Chair to the members of the Panel represented at the Session a list of Nominees for each office for which an election is to be held."[207] The rules of the IPCC left the nominations of the Bureau generally open to the choice of the governments, but that "it is highly desirable that governments ensure the scientific-technical integrity and credibility of the IPCC" by nominating experts in the climate field.[208] A lengthy discussion period followed the Nineteenth Session over the election rules. Draft procedures for Bureau elections began with a document from the Vice-Chair, Richard Odingo of Kenya, at the 21st Session[209] (November 2003). The contentious issue was passed on to a formal task group under Odingo and David Warrilow of the UK, and more drafts followed.[210] Not until the 25th Session were new measures for election adopted.[211]

The new rules were first utilized at the 29th Session (September 2008), which was the site of the most interesting election results in the history of the IPC. As usual, many of the Bureau positions were by acclamation. The only two positions that had to be voted upon were the Co-Chair for WG I (developed nation) and for WG III (developing nation). The latter election was especially tight as Ramón Pichs Madruga of Cuba received 69 votes and Youba Sokona of Mali 68 votes. The IPCC decided to accept both members as Co-Chairs.[212] Again there was only one candidate nominated (by India) by the member states for the position of Chairman, Pachauri.

The re-election of Pachauri by acclamation appeared to confuse some delegates. Comments were once more made about the "clarity" of the Rules of Procedures.[213]

Yet again, the election rules were re-examined. At the following three Plenary Sessions, new recommendations were circulated.[214] A major change, upon a recommendation of the InterAcademy Council (IAC) inquiry in 2010,[215] was that the position of Chairman (and Chairs and Vice-Chairs of the Working Groups) would be limited to one term.[216]

Structure & rules

The IPCC was to supply science for policy, but policy for science was lacking within the IPCC. Bolin can be criticized for allowing a lack of formal controls within the IPCC but Bolin was a leading climate scientist and scientific endeavors generally did not follow such formal regulations. The lack of standard operating procedures and protocols originated from the nature of the beast – science should not be controlled by rules:[217]

> The IPCC is run by scientists. Its participants think of it primarily as a scientific body. By the standards of many political organizations, its formal rules are not very extensive …
>
> There are good reasons for this arrangement. Formal procedures are relatively unimportant in scientific culture. This is true because scientists belong to very small social groups endowed with extremely strong and deeply entrenched (informal) norms. In addition, since scientific methods and results are constantly changing, too much focus on formal rules would inhibit progress. Likewise, formal rules are not very important in the day-to-day functioning of the IPCC. Instead, "informal rules based on the everyday practices of scientific communities" guide the bulk of the work.

But just before the same authors also stated: "[T]he IPCC needs better, more explicit rules of procedure."[218] Bolstered by some of the world's leading climate scholars, there was little doubt that they were well equipped to handle the science without being told what to do. While true, the IPCC was no simple scientific body. And, despite the haughty claims, there were fundamental reasons why standard operating procedures should have been important for the Assessment Report process:

- The IPCC was not carrying out a review for scientists; the customers (i.e., governments) had strict demands on how this was to be carried out.
- While those involved were generally good at science, they were not always good editors or leaders, nor so good with the politics of the IPCC.
- With less experienced members being added, especially from developing countries, more exact rules would have aided the newcomers.
- The intense public scrutiny was new to the scientific world. Scientists were used to observing through a microscope, but it was now science that was being observed. Any little flaw could damage the integrity of the Assessment Reports or even threaten the entire IPCC.

At the beginning of the IPCC, the process had been managed in an *ad hoc* manner. No documented rules or procedures were issued in the first three years; and then the first *Principles Governing IPCC Work* consisted of just 12 brief paragraphs, which covered:[219]

1. IPCC tasks would come from WMO and UNEP.
2. The IPCC Bureau would reflect a balanced geographical distribution.
3. The IPCC Working Groups should have clearly defined and approved mandates.
4. The Chairman should invite governments and other bodies to the IPCC.
5. Governments should be informed in advance of invitations of experts to join the IPCC.
6. The IPCC should do its best to reach consensus.
7. The conclusions of the Working Groups are not official IPCC views until accepted by the IPCC.
8. Invitations should be at least six weeks in advance.
9. Documents should be sent to the IPCC Secretariat at least four weeks in advance.
10. Interpretation in all languages should be available at sessions.
11. The timing of sessions of the Working Groups should be coordinated with the meetings of the INC.
12. Principles will be reviewed annually.

Later, Bolin commented on the abbreviated list:[220]

> It is quite remarkable that the rules of procedure were simply a dozen short paragraphs. This undoubtedly implied that the IPCC had gained the confidence of the countries that had played a leading role in the IPCC so far. This was going to be most valuable in the future.

Procedures for the Assessment Reports had been updated at the Ninth Session of the IPCC[221] (June 1993), and revised Principles Governing IPCC Work was approved at the Fourteenth Session in 1998.[222] Among the major changes:

1. The UNFCCC was added as a source of tasks for the IPCC.
2. "Review is an essential part of the IPCC process."
3. The Principles should be reviewed at least every five years.
4. A procedure for Reports was added as an Appendix.
5. Financial procedures were added as an Appendix.

In 2003 (21st Session), Pachauri assigned a task force to review them.[223] As expected, the task force recommended that there was no need for revisions of the Principles.[224]

What jumps out at the external observer is the lack of detail on how these great intentions are supposed to materialize. Simply issuing a bunch of principles does not ensure that they are being followed, even when the intended audience includes eminent scholars and Nobel Prize winners. The rules of the IPCC were too casual for the immense task at hand. And many of those participating in the preparation of the Assessment Reports and members of the IPCC, itself, later stated that the process was "poorly understood":[225]

> From extensive oral and written input gathered by the Committee, it is clear that several stages of the assessment process are poorly understood, even to many scientists and government representatives who participate in the process. Most important are the absence of criteria for selecting key participants in the assessment process and the lack of documentation for selecting what scientific and technical information is assessed.

Increasing concern was raised within the IPCC about the *laissez-faire* attitude that the IPCC was operating under. At the 28th Session member states and authors were invited to suggest improvements to the structure and procedures of the IPCC.[226] In the *Future of the IPCC*, over 140 pages of comments were recorded.[227] Not until the 30th Session in April 2009 were formal recommendations made to the IPCC.[228] But it was too little, too late. In but a few months, the IPCC was under attack.

A scandal led to the IAC inquiry which recommended several major changes to the procedures of the IPCC, including author selection, uncertainty, gray literature, the review process, preparation of the SPM and errors in the issued Assessment Report; many of which were accepted in principle at the 32nd Session[229] (October 2010). Task Groups were formed by the IPCC to deal with the details[230] and changes were approved in later Sessions.[231] I will return to the IAC inquiry later (see Chapter 2: "Foul ball: the 2010 season").

Politicization of science

A misconception is that the IPCC is a way of depoliticizing the climate change issue; far from it, the IPCC invites politicization into the scientific review,[232] but under controlled circumstances. The "membership" of the IPCC is often confused as the scientists preparing the Assessment Reports are not (usually) official members of the IPCC; membership is restricted to nominations from a country to represent a national government. A high-profile example took place at the Sixth Session in 1991, when Robert Reinstein had replaced Fred Bernthal, as Chair of Working Group III.[233] The exchange of American Chairs demonstrated the nation-orientation of the appointments. George H. W. Bush had become the new US President, and he simply wanted his "scientist" to represent the nation.[234] The same thing then happened with Reinstein, when President Clinton appointed Robert Watson to the IPCC.[235]

The IPPC had thus been appointed by the United Nations to carry out scientific assessments by volunteer scholars, but under national political control. While scientists dominate the input, politicians then framed the picture painted by the scientists. Traditionally, politics and science are like oil and water: they do not mix well. The IPCC has thrown them into the same pot, as they "interplay between science and politics."[236] The two groups are suspicious of each other, each one fearing that the other will distort the message; the omniscient scholars in their ivory towers were expected to exaggerate the seriousness of the issue and advocate policies contrary to the best interests of their nations – technocracy could triumph over democracy; from the other side, the Machiavellian bureaucrats were expected to censor the scientific-based facts of the perils of climate change, so that national economic interests would be protected.

There are firm rules in the IPCC to prevent scientists from being policy prescriptive. The credo of the IPCC is: scientific integrity, policy relevant and policy neutral. Obasi of the WMO, and Tolba of UNEP, would often welcome the delegates to the early Plenary Sessions of the IPCC, and the two senior UN officials would freely give their opinions on the science of climate change and what should be done.[237] Since the IPCC itself was to determine the state of the science and recommendations, such comments by the directors of the WMO and UNEP could have been interpreted as interference. At the inaugural meeting of the IPCC held in Geneva on November 9, 1988, Tolba even stated that the Panel "should bravely inform the world what ought to be done." [238]

However, this was strictly against their rules. Tolba later stated that this was not the role of the IPCC.[239] The mandate of the new organization required them to be "neutral with respect to policy."[240]

Though hard rules did not provide similar protection for the science from the politicians, scientific integrity was also not supposed to be at risk. Bert Bolin summarized the situation:[241]

> Scientists as well as politicians need to recognize their different roles. The former must protect their scientific integrity, but also respect the role of politicians ... The Assessment process must involve the very best scientists ... Scientists need to inform politicians in a simple manner that can be readily understood, but the message must always be scientifically exact.

The Chairman of the IPCC guarded against political excesses. This unwritten rule had been established during the tenure of Bolin, who had warned of the "danger that political value judgments may also penetrate inside the IPCC process."[242] He noted in the Plenary Sessions that this had taken place, but the Working Groups of the IPCC had been "well guarded." That even the Working Groups had to be "well guarded" to prevent politicization demonstrated how close at hand the threat was. The risk was continually there; for example:

- *Third Session* (February 1990) – an acknowledgement was made that the IPCC had "worked hard" to prevent "politicization" of science;[243]
- *Tenth Session* (November 1994) – the Chairman of the INC felt it necessary to confirm that the IPCC "remained unimpaired by political and economic pressures";[244]

- *28th Session* (April 2008) – there were fears that the "catering" to policymakers was affecting scientific objectivity.[245]

The Assessment Report system was most vulnerable to politicization at the Summary for Policymakers (SPM) meeting at the Plenary Session;[246] a member described the fears of such political interference (for a Special Report):[247]

> We all knew that much of the debate and rewriting of the Summary for Policy Makers would be based on a balance of the interests of the Parties to the Convention and the Protocol rather than from the sole perspective of maximizing the scientific integrity and clarity of the Report.
>
> But in fact that, even in those circumstances, after 7 days and nights, we've approved by consensus a balanced and objective Summary for Policy Makers and accepted, with minimum changes, an extremely comprehensive underlying report, is an enormous tribute to the professionalism, integrity and good will of many people.

At the Seventeenth Session (April 2001), the Chairman, Robert Watson, expressed similar views.[248]

During these discussions, politics, not science, dominated the discussion.[249] Pseudo-scientific debates ensued between those nations that wanted to highlight impacts of climate change, often from Europe, and those which wished to tone down the severity of the warnings, especially from the US. Watson had to intervene during the IPCC Plenary Session on the SPM of the Synthesis Report of the TAR:[250]

> Views often differed on how the findings of the working groups should be synthesized, interpreted and reflected, and on how to ensure consistency between the working group and Synthesis Report SPMs. Opinions also differed on the messages that should be conveyed to policymakers. Throughout the discussion, Chair Watson urged delegates not to alter text taken directly from the approved working group SPMs and to refrain from adding more detailed text.

For the AR4, over 1,500 comments, with more than half on the SPM, had been received.[251] Four full days of the Plenary Session were needed for the line-by-line approval of the SPM and only one day for the page-by-page approval of the much longer full Synthesis Report. In one of the discussions, the US representative (the "politician") wanted a sentence changed[252] and an author (the "scientist") disagreed. The Chairman of the IPCC, Pachauri, suggested that the sentence be deleted, which was supported by the US and China but opposed by France and Germany. Since consensus could not be reached, the Chairman again suggested removing the controversial sentence. Germany again disagreed, but the phrase was removed from the SPM.[253]

This practice of removing items of disagreement from the SPM cannot be considered protecting scientific integrity. Further, even when items were not removed, vital information could still be lost from the SPM. An insider within the IPCC stated that governments "succeeded in diluting many of the paragraphs."[254] Here we have the greatest weakness of the Assessment Report process. The science largely survived political manipulation, but the message could be downplayed. Politicization of the Assessment Reports is inevitable; the degree of this influence, though, appears to be relatively muted considering the circumstances.

The term "Summary" in Summary for Policymakers is a misnomer for if changes were made at the Plenary Session to the SPM, the main report would have been modified to take the "summary" changes into account. Since the few pages of SPM of the Synthesis Report may be the only section of the Assessment Reports that policymakers and governmental officials, especially high-level officials, will ever read, the message from this section is paramount to the entire process.

The grand finale of the IPCC process was the approval of the SPM of the Synthesis Report, which only happened every six years or so. Although often viewed as being redundant in the minds of the scientists, the SPM of the Synthesis Report was the most important part to policymakers.[255] Most Panel members had little interest in the dry technical details found in the hundreds of pages of the Assessment Reports. In this respect, the brief meeting of the SPM on the Synthesis Report is an anti-climactic (not to mention anti-climatic) end to the Assessment Report process. The hard work of preparing the comprehensive scientific review over the past few years had already been completed. The situation is sad in a way, as thousands of the leading scholars in the world labored over the 1,000 pages or so of the Assessment Report, and, in the end, the whole procedure basically came down to a handful of pages in the infamous SPM, whose final version was largely controlled by others.

Regional representation

Another type of politicization became ingrained into the IPCC, that in many ways was more disruptive than the agendas of the government members. The IPCC had another compulsory obligation – regional representation; the IPCC had to ensure that developing nations were fairly represented in all aspects of the IPCC work and structure. The external constraint came from being an intergovernmental body of the United Nations. There were more discussions in the minutes of the Plenary Sessions on geographic balance than on scientific integrity, politicization, or any other single issue, and the participation of the developing countries became the agent of change within the IPCC.

While all nations were formally welcomed to the IPCC,[256] many countries were limited, both from a financial and expert point of view (and some by language as the scientific literature was dominated by English articles).[257] These nations demanded that more effort be made to ensure their fuller participation within the IPCC.

Even before the First Plenary Session, regional participation was on Bolin's mind. As the IPCC was just getting organized in the summer of 1988, Bolin was asked if another review of climate change by a new group (i.e. the IPCC) would be worthwhile. His response was: "Right now, many countries, especially developing countries, simply don't trust assessments in which their scientists and policymakers have not participated ... Don't you think global credibility demands global representation?"[258]

Despite the words of Bolin, upon its inception in 1988, the IPCC faced the challenge of limited developing nation participation. Symptomatic of the problem was the site of the Plenary Sessions. Of the first seven sites only one was held in a developing country – the Second Session (June 1989) at the UNEP headquarters in Nairobi. Only after the Tenth Session (November 1994) were meeting sites more international in nature. A sense of exclusivity had been unintentionally introduced at the First Session by the designation of a limited number of "core members" to each Working Group: WG I–13, WG II–13 and WG III–17,[259] who, by the way, were chosen by the infamous Nominations Committee. The good intentions of limiting the numbers to more controllable levels were a barrier to membership in the Working Groups by developing countries. The core member quota was removed at the Second Session.[260] One frustrated member complained about the way they had been treated and that "it was felt to date that developing countries were being brought along to the IPCC to create the impression only of being part of the action."[261]

Obasi and Tolba had expressed their concern at the Second Session,[262] and, in subsequent Plenary Sessions, they had to constantly remind the IPCC that improvements had to be made to allow all countries to participate. The IPCC had to divert limited human resources to deal with the regional representation problem. In February 1989, at the First Session of the Bureau,[263] the Vice-Chairman of the IPCC, Al-Gain of Saudi Arabia, had been asked by the Bureau to "collaborate" with Brazil, Senegal and Zimbabwe on this task. The new group was called the *Ad-hoc* Sub-group on Ways to Increase the Participation of the Developing Countries in IPCC Activities. The Al-Gain *Ad-hoc* Sub-group reported back at the Second Session;[264] and the importance of the issue was raised up a notch when a more formal group was formed: a Special Committee on the Participation of the Developing Countries,[265] chaired by Jean Ripert of France (who later headed the INC), and ten members selected equally from developed and developing nations. However, the IPCC was not moving fast enough for some. Before the Committee had a chance to report back to the IPCC, complaints had been raised by some member states to the UN General Assembly,[266]

> Concerned that the participation of the developing countries in the Intergovernmental Panel on Climate Change remains limited, and stressing the need for the Intergovernmental Panel, in view of its intergovernmental nature to do all it can to ensure adequate participation and governmental involvement in its activities in accordance with United Nations practice.

During the Fourth Session in August 1990, Ripert's Special Committee formally recommended a number of financial and other measures to aid the participation of developing countries.[267] The Fourth Session should have been a festive affair as the First Assessment Report (FAR) was to be formally approved. However, Obasi acknowledged the tension during the meeting.[268] He addressed the IPCC and stated his "one regret" was with regard to the participation of developing nations;[269] he admitted that the short timetable for the FAR had resulted in some "interested parties" being excluded.[270] Even during the historic occasion of the approval of the FAR, regional participation dominated the mood of the meeting.

Illustrating the significance of this non-scientific issue, the work of the Special Committee actually appeared in the FAR. Obasi and Tolba mentioned it in the short *Preface*, and the recommendations of the Special Committee had a page in the *Overview*,[271] and in the *Forward* to the FAR of WG II; the Chairman of Working Group II stated that a "lack of sufficient funding"[272] had prevented greater participation by the developing countries. One would have thought that discussion of this internal organizational matter would be inappropriate in the First Assessment Report (FAR), one of the most important *scientific* documents in history.

Discontent within the IPCC continued to grow. A group of developing nations had met on February 7, 1990, to discuss the problem. Brazil and Mexico were especially unhappy. In September 1990 they raised the subject to the WMO and UNEP, and again to the General Assembly of the UN. The actions by these nations led the UN to remove responsibility for the climate negotiations away from WMO and UNEP, and to establish the INC (*vide supra*).[273] At the Fifth Session in March 1991, Tolba again scolded the IPCC and acknowledged the "frustration of some developing countries."[274] He explained that these nations expected to be equal partners in the INC negotiations and that the impacts and responses to climate change for their countries were best assessed by representatives from their own regions. And Obasi stressed that the IPCC would be more effective if developing countries were more active within the Panel.[275] During this session, the *Principles Governing IPCC Work* was adopted for future reports (after the FAR), [276] which stated that the IPCC Bureau and Working Groups "shall reflect balanced geographic representation." While recognized, no reorganization of the IPCC was undertaken to accommodate the new rules.[277]

While Obasi and Tolba were great at complaining, they were short of answers and did not offer any extra financial assistance. Bolin's patience with this political diversion from the main role of the IPCC came to an end; he put the responsibility back on the WMO and UNEP by stating that the IPCC needed more funding from them to do something.[278] Financing was important; of the 53 developing countries that attended the Sixth Session, 50 received support from the IPCC Trust Fund.[279] The solution was largely, but not completely, financial; something the IPCC was not set up to deal with, even though it tried to help. Through the early years of the IPCC their financing was limited and somewhat arbitrary. The long-term financial structure of the IPCC was, frankly, a mess.[280] The Trust Fund was based on the voluntary donations of member states. Several times the IPCC almost ran out of funds.[281] Although there was continuous pressure for the IPCC to foster the participation of developing countries, funding for such an expense was not available. If this was so important to the WMO and UNEP, they should have handled it directly or, at least, have removed the financial burden from the IPCC.

Tolba had rather harsh words in his written statement to the Sixth Session (October 1991), reminding members that he had raised this pressing issue back at the Third Session[282] (February 1990). Since the IPCC had not made any progress, he would tell them what to do. Tolba provided recommendations on how this should be accomplished by expanding the membership of the Bureau to 30 members, divided evenly between developed and developing nations:[283]

- Two Vice-Chairmen positions being added
- Each Working Group should have one Chairman, two Vice-Chairmen and one Rapporteur.

The new Bureau would also include special regional representatives from:

- Africa
- Australia and Pacific
- Latin America and the Caribbean
- Europe
- Others.

Tolba implied that he was trying to protect the IPCC as there were forces at work trying to discredit it over the issue of geographical representation:[284]

> [M]y main consideration is that the sterling job accomplished by the Panel should not be jeopardized by possible dissatisfaction from any of the actors. It is my view also that when taking these decisions, the scientific integrity of the Panel will not be compromised.

His latter comment implies that he had received feedback that there was concern that affirmative action with regard to regional diversification would hurt the scientific integrity of the IPCC.

Bolin had always wanted the best scientists to be involved with the IPCC, and he was not simply going to increase the role of developing countries, if such actions placed the science at risk. At the same time, he knew that he had to do something. This crisis could potentially tear the IPCC apart. In a somewhat defiant response, and illustrating the independence of the IPCC, Bolin agreed that the role of the developing countries should be increased within the Bureau, but how to do this was up to the Panel,[285] and that the "scientific integrity and rigor of the Panel's work be preserved."[286]

The Panel did agree to an interim measure that five additional Vice-Chairs be added to the Working Groups to improve geographical representation.[287] Of the five, one went to Germany. The IPCC set up a new Task Force under Bolin to look at the structure of the Bureau. First, the Task Force on IPCC Structure was to establish a structure to best fulfill its scientific and technical assessments. Fair geographical representation was only the fourth item listed. At the Eighth Session (November 1992), which followed the Earth Summit in Rio, the IPCC approved

Table 1.2. Country participation in the plenary sessions of the IPCC.[314]

IPCC	Non-OECD	Total Countries
First Session (November 1988)	47%	30
Second Session	57%	44
Fourth Session	69%	70
Sixth Session	72%	79
Eighth Session	80%	96
Ninth Session	82%	115
Eleventh Session (December 1995)	84%	117

Source: Agrawala, Structural and Process History 1998, p. 630; for an overview of the developing nation situation at the IPCC until this time see Agrawala, Structural and Process History, 1998, pp. 628–32.

the recommendations.[288] The Bureau was restructured along similar lines to the suggestion by Tolba, and was increased to 28 members, structured thus:[289]

- Two Vice-Chairs, of which, at least one would be from a developing country
- Three Working Groups
  1. Science – Two Chairs and Two Vice-Chairs
  2. Impacts, response options and mitigation – Two Chairs and Eight Vice-Chairs (two for each sub-group)
  3. Cross-cutting economic issues – Two Chairs and Two Vice-Chairs
- Regional representatives from: Africa, Asia, Europe, North and Central America, South America, and South-West Pacific
- Term of all Chairs, Vice-Chairs and Bureau members was three years.

Change was also made to the Assessment Report process. The Bureau had agreed that at least one Lead Author of the Assessment Reports would be from a developing country.[290] Bolin must have been reluctant about supporting this latter rule; a factor other than scientific expertise would now have to be considered when choosing the writing teams.

Obasi announced that the WMO endorsed the restructuring of the IPCC, but he still stressed the importance of the full participation of the developing countries in the activities of the IPCC.[291] While still mentioned, Obasi's message on this sensitive topic was less severe by the time of the Tenth Session[292] (November 1994). The changes translated into results as the participation of developing countries steadily improved (Table 1.2).

Since the major reorganization, the Bureau membership has been tweaked, but no major changes have taken place. At the last Session under Bolin, the Bureau increased its membership to 30,[293] which was double the original size a decade earlier. And at the Eighteenth Session (September 2001), there had been a proposal to add another 20 members because of the workload – but the recommendation was fortunately rejected. Watson responded that the real problem was that some members of the Bureau were not carrying out their duties, especially among the

Table 1.3. Evolution of the IPCC Bureau.

Session	Chair	Vice-Chair	Working Group		Other	Total
			Chair	Vice-Chair		
1st (1988)	1	1	3	9	1 (Rapporteur)	15
8th (1992)	1	2	6	13	6 (Regional Reps.)	28
13th (1997)	1	5	6	18		30
18th (2001)	1	3	6	18	2 (Task Force Chairs)	30
35th (2012)	1	3	6	19	2	31

regional members. He suggested that the regional groups should set up a mechanism to oust such inactive representatives.[294] Within diplomatic circles these were harsh words, especially coming from the Chairman. A summary of the evolution of the structure of the Bureau is presented in Table 1.3.

The issue seemed to have died down[295] until 2009, when the pressure switched from the structure of the Bureau to the writing teams of the Assessment Reports.[296] The metrics spoke for themselves. In the Fourth Assessment Report (AR4), 67 percent of the authors were from developed countries. Among the reviewers, the share was more disproportionate, at 82 percent.[297] There is no doubt that the authors, experts and reviewers are dominated by developed nations,[298] and the Assessment Reports would have had greater global credibility if there had been broader representation. Criticism in this regard abounds in the literature,[299] but how is the IPCC to change the scientific expertise distribution across the globe? The argument can be made that science is borderless, but the reality is that it is carried out where the money is available to do such work, and most climatic scholars, including those originally from the developing world, were working in the developed world. Making matters worse, the metrics adopted by the IPCC to count regional representation minimized the counting of developing country experts. The data were determined only by the country where the expert was working, not the country of origin of the scientist. If an expert from a developing country was working in a developed nation, which often happened, the scientist counted as being from a developed country.

Members of the developing nations understood the problems restricting their participation:[300]

A number of individuals who were interviewed or responded to the Committee's questionnaire observed that developing country scientists often had limited understanding of developments outside of their region and/or did not do (or were not asked to do) their fair share of the work. Most attributed this lack of participation to the unique difficulties faced by developing-

country scientists. These include the exclusive use of English to communi-
cate during the preparation of the Working Group reports, the lack of support
by their home institutions, poor access to literature, and the relatively small
number of qualified scientists from some developing countries.

It was not clear if the developing countries were even worried about their level of
participation in the Assessment Report process. In a questionnaire on the matter
by the IPCC, only 18 developing countries even bothered to respond.[301] Developed
country participation, on the other hand, was proportionally much higher. Who
was really concerned about this alleged problem?

By definition, geographical representation is a political not scientific matter.
Politics interfering with the science of the IPCC had long been a threat, but major
interference by governments in the reports of the IPCC had largely been controlled. With regional affirmative action, political interference became embedded
in the organization. This represented the politicization of science at the highest
level – demanded by the UN itself.

While, in principle, the developing countries should have been more represented, this forced a dilemma. Setting objectives – scientific integrity and regional
representation – which potentially conflict with each other, is a recipe for failure.
A troubling aspect is that the nagging pressure from the WMO and UNEP implied
that regional representation was to take priority. Although reluctant to do so, the
IPCC gave way to the demands of their parent organizations, even though quotas
on regional representation could only cause a dilution in expertise; the good intentions inherently resulted in the best experts not necessarily leading the IPCC or
the preparation of the Assessment Reports. Bolin candidly admitted this problem
with the developing country Chairs of the Working Groups: "[T]hose from developing countries brought other [than 'very knowledgeable in the environmental
sciences'] but more limited expertise, and their participation was essential as an
education process."[302]

The prime mandate of the IPCC was to prepare the Assessment Reports by the
leading scientists, and the organization was distracted from the main task at hand
and committed limited resources to what must be best described as a peripheral
issue. More than that, the members of the IPCC seemed to have become lost
within this discussion, as they forgot, or were forced to forget, why the IPCC was
formed in the first place (it was not for capacity-building in developing countries).
Certainly, the IPCC would have been better off if the passion spent on regional
representation had been directed towards the Assessment Report process.

Considering the immense undertaking of reviewing the science of climate
change and the tight deadlines, the diversion of scarce resources to this non-scientific issue highlighted its importance, but was it made too important? The weaken-
ing of scientific integrity by regional representation is difficult to assess and one
must be careful when assigning blame. Is there evidence that the IPCC was being
harmed by regional representation? Even within the ranks of the overly politically
correct IPCC, concern was raised by the Co-Chairs of WG III in the preparation
of the TAR:[303]

The main problem was the non- or inadequate performance of some LAs and CLAs, indicating that the selection process was not perfect. Because of the sensitivity of the regional balance of writing teams only in a few cases LAs were dropped or replaced. In particular for developing country LAs, availability of time and lack of internal budgets to cover time spent led to less than optimal performance in the writing teams.

Although cushioned with political etiquette, the blame is laid on regional representation. Later, Jeff Price, an LA of TAR and AR4, stated that the best experts should be chosen and not be influenced by gender and geographical location.[304]

The IPCC had entered a dangerous zone by interfering with the Assessment Report process. It was risky enough that regional representation was allowed to meddle with the structure of the Bureau. When applied to the assessment process itself, this threatened what Bolin had so diligently protected during his long tenure – scientific integrity.[305] During the preparation of the Fourth Assessment Report (AR4), a scandal, known as "glaciergate," would emerge that would put into question the scientific integrity of the IPCC. While regional diversification cannot be held responsible for this failure of the Assessment Report system, its contribution to it cannot be simply ignored either.

References[306]

Agrawala, S., Context and Early Origins of the Intergovernmental Panel on Climate Change, *Climatic Change* 39 (4), 605, 1998; also see *Explaining the Evolution of the IPCC Structure and Process.* ENRP Discussion Paper E-97–05, Kennedy School of Government, Harvard University, August 1997; http://www.hks.harvard.edu/gea/pubs/e-97-05.pdf

Agrawala, S., Structural and Process History of the Intergovernmental Panel on Climate Change, *Climatic Change* 39 (4), 621, 1998

Bodansky, D., The Framework Convention/Protocol Approach, 1999; http://whqlibdoc.who.int/hq/1999/WHO_NCD_TFI_99.1.pdf

Bolin, B., *A History of the Science and Politics of Climate Change*, Cambridge University Press, 2007

Bolin, B., Politics and the IPCC, *Science* 296 (5571), 1235, 2002

Bolin, B., Science and Policymaking, *Ambio* 23 (1), 25, 1994

Bose, I., 'We Cannot Sell the Idea of Equity Back Home,' says U.S. Negotiator, *Down to Earth*, November 29, 2012; http://www.downtoearth.org.in/content/we-cannot-sell-idea-equity-back-home-says-us-negotiator

Bruce, J.P., Intergovernmental Panel on Climate Change and the Role of Science in Policy, *ISUMA* (winter), 11, 2001

Brundtland Commission's Report, *Report of the World Commission on Environment and Development: Our Common Future*, 1987; http://www.un-documents.net/wced-ocf.htm

Cass, L.R., *The Failures of American and European Climate Policy*, SUNY, 2006

Cole, D.H., Climate Change and Collective Action, May 2008; http://papers.ssrn.com/sol3/papers.cfm?abstract_id=1069906

Dales, J.H., *Pollution, Property & Prices*, Edward Elgar, 2002

Davies, D.A., The Role of WMO in Environmental Issues, *International Organization* 26, 327, 1972

Earth Negotiations Bulletin; http://www.iisd.ca/vol12/

Edwards, P.N., and Schneider, S.H., Broad Consensus or "Scientific Cleansing," *Ecofable/Ecoscience* 1 (1), p. 3, 1997; http://pne.people.si.umich.edu/PDF/ecofables.pdf

EPA, Clean Air Markets, Emissions, Compliance and Market Data; http://www.epa.gov/airmarkets/progress/ARP09_1.html

European Commission, *The Greenhouse Effect and the Community, Communication*, November 1988; http://aei.pitt.edu/5684/01/003076_1.pdf

Fleming, J.R, *Historical Perspectives on Climate Change*, Oxford University Press, 1998

Franz, W.E., IIASA, *The Development of an International Agenda for Climate Change: Connecting Science to Policy, Interim Report IR-97–034/September*, September 1997; http://www.iiasa.ac.at/Publications/Documents/IR-97-034.pdf

Hecht, M.T., and Tirpak, D., Framework Agreement on Climate Change: A Scientific and Policy History, *Climatic Change* 29 (4), 371, 1995

Hevesi, D., Bert Bolin, 82, Is Dead; Led U.N. Climate Panel, *New York Times*, January 4, 2008; http://www.nytimes.com/2008/01/04/obituaries/04bolin.html

Hilsenrath, J., Cap and Trade's Unlikely Critics: Its Creators, *Wall Street Journal*, August 13, 2009

Hulme, M., and Mahony, M., Climate Change: What do We Know About the IPCC?, *Progress in Physical Geography*, 34 (5), 705, 2010

InterAcademy Council, *Climate Change Assessments, Review of the Processes and Procedures of the IPCC*, 2010; http://www.interacademycouncil.net/24026/26050.aspx

IPCC, *First Assessment Report*,[307] *Overview*, 1990

IPCC, *First Assessment Report, Working Group I: Scientific Assessment of Climate Change*, 1990

IPCC, *IPCC Statement on New Scientist Article*, Statements 311, June 24, 2012

IPCC, *First Assessment Report, Working Group II: Impacts Assessment of Climate Change*, 1990

IPCC, *Principles and Procedures of the IPCC*; http://ipcc.ch/organization/organization_procedures.shtml

IPCC, *Procedures for the Preparation, Review, Acceptance, Adoption, Approval and Publication of IPCC Reports*, April 1999; http://www.ipcc.ch/pdf/ipcc-principles/ipcc-principles-appendix-a.pdf

IPCC, *Report of the First Session*[308] *of WMO/UNEP Intergovernmental Panel on Climate Change*, Geneva, November 9–11, 1988

IPCC, *Report of the Second Session of WMO/UNEP Intergovernmental Panel on Climate Change*, Nairobi, June 28–30, 1989

IPCC, *Report of the Third Session of WMO/UNEP Intergovernmental Panel on Climate Change*, Washington, DC, February 5–7, 1990

IPCC, *Report of the Fourth Session of WMO/UNEP Intergovernmental Panel on Climate Change*, Sundsvall, Sweden, August 27–30, 1990

IPCC, *Report of the Fifth Session of WMO/UNEP Intergovernmental Panel on Climate Change*, Geneva, March 13–15, 1991

IPCC, *Report of the Sixth Session of the Intergovernmental Panel on Climate Change*, Geneva, October 29–31, 1991

IPCC, *Report of the Eighth Session of the Intergovernmental Panel on Climate Change*, Harare, Zimbabwe, November 11–13, 1992

IPCC, *Report of the Ninth Session of the Intergovernmental Panel on Climate Change*, Geneva, June 29–30, 1993

IPCC, *Report of the Tenth Session of the Intergovernmental Panel on Climate Change*, Nairobi, November 10–12, 1994

IPCC, *Report of the Eleventh Session of the Intergovernmental Panel on Climate Change*, Rome, December 11–15, 1995

IPCC, *Report of the Twelfth Session of the Intergovernmental Panel on Climate Change*, Mexico City, September 11–13, 1996

IPCC, *Report of the Thirteenth Session of the Intergovernmental Panel on Climate Change*, Maldives, September 22 & 25–28, 1997

IPCC, *Report of the Fourteenth Session of the Intergovernmental Panel on Climate Change*, Vienna, October 1–3, 1998

IPCC, *Report of the Seventeenth Session of the Intergovernmental Panel on Climate Change, Future Work Programme of the IPCC*, Nairobi, April 4–6, 2001

IPCC, *Report of the Eighteenth Session of the Intergovernmental Panel on Climate Change, Future of the Intergovernmental Panel on Climate Change*, Wembley UK, September 24–29, 2001

IPCC, *Report of the Nineteenth Session of the Intergovernmental Panel on Climate Change*, Geneva, April 17–20, 2002

IPCC, *Report of the Nineteenth Session of the Intergovernmental Panel on Climate Change, Evaluation of the Work of Working Group III and Its Technical Support Unit from 1998–2002: Lessons for the Future*, April 2002

IPCC, *Report of the Nineteenth Session of the Intergovernmental Panel on Climate Change, IPCC Programme and Budget for 2003–2005*, April 2002

IPCC, *Report of the Nineteenth Session of the Intergovernmental Panel on Climate Change, Options for the IPCC Budget Management*, December 2001

IPCC, *Report of the Twentieth Session of the Intergovernmental Panel on Climate Change*, Paris, February 19–21, 2003

IPCC, *Report of the Twentieth Session of the Intergovernmental Panel on Climate Change, Preparations for the Fourth Assessment Report*, February 2003

IPCC, *Report of the 21st Session of the IPCC*, Vienna, November 3 & 6–7, 2003

IPCC, *Report of the 21st Session of the IPCC, A Proposal for the Synthesis Report (SYR) for the AR4*, November 2003

IPCC, *Report of the 21st Session of the IPCC, Election Procedures for the IPCC Bureau and TFB*, November 2003

IPCC, *Report of the 21st Session of the IPCC, Review of IPCC Principles*, November 2003

IPCC, *Report of the 22nd Session of the IPCC, New Delhi*, November 9–11, 2004

IPCC, *Report of the 22nd Session of the IPCC, Rules and Procedures for the Election of the IPCC Bureau and Task Force Bureau*, November 2004

IPCC, *Report of the 24th Session of the IPCC*, Montreal, September 26–28, 2005

IPCC, *Report of the 24th Session of the IPCC, Revised Draft Rules and Procedures (ROP) for the Election of the IPCC Bureau and Task Force Bureau*, September 2005

IPCC, *Report of the 25th Session of the IPCC*, Port Louis, Mauritius, April 26–28, 2006

IPCC, *Report of the 25th Session of the IPCC, Draft Rules of Procedures for the Election of the IPCC Bureau and Any Task Force Bureau*, April 2006

IPCC, *Report of the 25th Session of the IPCC, Review of IPCC Terms of Reference*, April 2006

IPCC, *Report of the 26th Session of the IPCC, Review of IPCC Terms of Reference*, May 2007

IPCC, *Report of the 26th Session of the IPCC, Review of IPCC Terms of Reference, Compilation of Comments Received on the Draft Proposal*, May 2007

IPCC, *Report of the 26th Session of the IPCC, Review of IPCC Terms of Reference, Summary of Relevant Resolutions and Decisions*, May 2007

IPCC, *Report of the 27th Session of the IPCC, IPCC Fourth Assessment Report "Climate Change 2007" Draft Synthesis Report*, November 2007

IPCC, *Report of the 27th Session of the IPCC Bureau*,[309] Geneva, August 7–8, 2002; http://www.ipcc.ch/meeting_documentation/meeting_documentation_sessions_of_the_ipcc_bureau.shtml

IPCC, *Report of the 28th Session of the IPCC*, Budapest, April 9–10, 2008

IPCC, *Report of the 28th Session of the IPCC, Future of the IPCC, Comments Received after the Deadlines and Translations into English of Earlier Submissions*, April 2008

IPCC, *Report of the 28th Session of the IPCC, Future of the IPCC, Compiled Comments from Governments, Authors, Organizations and Bureau Members*, April 2008

IPCC, *Report of the 28th Session of the IPCC, Future of the IPCC, Synthesis of Comments*, April 2008

IPCC, *Report of the 29th Session of the IPCC*, Geneva, August 31– September 4, 2008

IPCC, *Report of the 30th Session of the IPCC*, Antalya, Turkey, April 21–23, 2009

IPCC, *Report of the 30th Session of the IPCC, Revision of the Rules of Procedures for the Election of the IPCC Bureau and Any Task Force Bureau*, April 2009

IPCC, *Report of the 31st Session of the IPCC , Implementation of Decisions Taken by IPCC-30 , Involving Developing/EIT Country Scientists*, Bali, October 26–29, 2009

IPCC, *Report of the 31st Session of the IPCC, Improving Participation of Developing/EIT Countries in the IPCC*, October 2009

IPCC, *Report of the 31st Session of the IPCC, Revision of the Rules of Procedures for the Election of the IPCC Bureau and Any Task Force Bureau*, October 2009

IPCC, *Report of the 31st Session of the IPCC, Revision of the Rules of Procedures for the Election of the IPCC Bureau and Any Task Force Bureau, Task Group on Rules of Procedures for the Election of the IPCC Bureau and Any Task Force Bureau*, October 2009

IPCC, *Draft Report of the 32nd Session of the IPCC*,[310] Busan, Republic of Korea, October 11–14, 2010

IPCC, *Report of the 32nd Session of the IPCC, Rules of Procedures for the Election of the IPCC Bureau and Any Task Force Bureau*, August 30, 2010

IPCC, *Report of the 32nd Session of the IPCC, The IPCC Fifth Assessment Report (AR5), Progress Report of the Cross Working Group Consultation on Article 2 of the UNFCCC*, October 2010

IPCC, *Report of the 33rd Session of the IPCC, Decisions Taken with Respect to the Review of IPCC Processes and Procedures, Governance and Management*, May 10–13, 2011

IPCC, *Report of the 33rd Session of the IPCC, Decisions Taken with Respect to the Review of IPCC Processes and Procedures, Procedures*, May 10–13, 2011

IPCC, *Report of the 33rd Session of the IPCC, Review of the IPCC Processes and Procedures, Proposal by the Task Group on Procedures*, May 10–13, 2011

IPCC, *Report of the 33rd Session of the IPCC, Review of the IPCC Processes and Procedures, Draft Proposals by the Task Group on Governance and Management, Terms of Reference for the Bureau and Bureau Members*, May 10–13, 2011

IPCC, *Report of the 34th Session of the IPCC, Further Work to Adopting Revisions to "Appendix C of the Principles Governing IPCC Work: Rules of Procedures for the Election of the IPCC Bureau and Any Task Force Bureau,"* November 18–19, 2011

IPCC, *Report of the 34th Session of the IPCC, Review of IPCC Processes and Procedures, Procedures: Adoption of the revised "Appendix A to the Principles Governing IPCC Work: Procedures for the Preparation, Review, Acceptance, Adoption, Approval and Publications of IPCC Reports,"* November 18–19, 2011

IPCC, *Report of the 35th Session of the IPCC, Procedures*, Appendix C, June 6–9, 2012

IPCC, *Report of the 35th Session of the IPCC, Review of the IPCC Processes and Procedures, Remaining Issues Related to Governance and Management, Comments by the UNEP Secretariat and WMO on the Draft Terms of Reference of the IPCC Secretariat and the IPCC Technical Support Units Contained in Document IPCC-XXXV/Doc.12*, June 6–9, 2012

IPCC, *Report of the 35th Session of the IPCC, Review of the IPCC Processes and Procedures, Remaining Issues Related to Governance and Management, Memorandum of Understanding between UNEP and WMO on the IPCC*, June 6–9, 2012

IPCC, *Request for AR4 Nominations*; http://www.ipcc.ch/meeting_documentation/ meeting_documentation_miscellaneous_correspondence.shtml

IPCC, *Second Assessment Report, Working Group I: The Science of Climate Change*, 1995

IPCC, *Third Assessment Report, Working Group I: The Scientific Basis*, 2001

InterAcademy Council, Renate, C., *IPCC Products, Procedures, and Processes, Secretary of the IPCC Renate Christ Presentation to the IAC Review Committee*, May 10, 2010; http://reviewipcc.interacademycouncil.net/

Jung, W., Expert Advice in Global Environmental Decision Making: How Close Should Science and Policy Get?, Belfer Center for Science & International Affairs, September 1999; http://www.hks.harvard.edu/gea/pubs/e-99-14.pdf

Karlsson, S., Srebotnjak, T., and Gonzales, P., Understanding the North–South Knowledge Divide and Its Implications for Policy: A Quantitative Analysis of the Generation of Scientific Knowledge in the Environmental Sciences, *Environmental Science & Policy*, 668, 2007

Leemans, R., Personal Experiences with the Governance of the Policy-Relevant IPCC and Millennium Ecosystems Assessment, *Global Environmental Change* 18, 12, 2008

McCauley, K., Barron, B., and Coleman, M., Crossing the Aisle to Cleaner Air, 2008; http://www.iop.pitt.edu/documents/casestudies/Crossing%20the%20Aisle%20to%20 Cleaner%20Air.pdf

Mintzer, I.M., and Leonard, J.A., (eds.), *Negotiating Climate Change*, Cambridge University Press, 1994

National Research Council, National Academy of Sciences, *Halocarbons: Effects of Stratospheric Ozone*, 1976

Netherlands Environmental Assessment Agency, Assessing an IPCC Assessment: An Analysis of Statements on Projected Regional Impacts in the 2007 Report, 2010; http:// www.pbl.nl/en/publications/2010/Assessing-an-IPCC-assessment.-An-analysis-of-statements-on-projected-regional-impacts-in-the-2007-report

Pachauri, R., IAC Review of the IPCC, Committee Meeting 1 – Public Session, May 14, 2010; http://reviewipcc.interacademycouncil.net/

Parks, B.C., and Roberts, J.T., Climate Change, Social Theory and Justice, *Theory, Culture & Society* 27, 134, 2010

Pearce, F., Climate Panel Adopts Controversial Grey Evidence, *New Scientist*, June 26, 2012

Pearce, F., Frankenstein Syndrome Hits Climate Treaty, *New Scientist*, 142 (1929), 5, June 11, 1994

Pearce, F., The Week the Climate Changed, *New Scientist*, 52, October 15, 2005

Pooley, E., *The Climate War*, Hyperion, 2010

Price, J., IPCC: Cherish It, Tweak It, or Scrap It, *Nature* 463, 732, February 11, 2010

Ramus, C.A., The Large Combustion Plant Directive: An Analysis of European Environmental Policy, 1991; http://www.oxfordenergy.org/?s=The+Large+Combustion +Plant+Directive%3A++An+Analysis+of+European+Environmental+Policy

Ravindranath, N.H., IPCC: Accomplishments, Controversies and Challenges, *Current Science* (Bangalore), 99 (1), 26, 2010

Renate, C., *IPCC Products, Procedures, and Processes, Secretary of the IPCC Renate Christ Presentation to the IAC Review Committee*, May 10, 2010; http://reviewipcc. interacademycouncil.net/

Russell, M., Boulton, G., Clarke, P., Eyton, D., and Norton, J., *The Independent Climate Change eMail Review*, July 2010; http://www.cce-review.org/pdf/FINAL%20REPORT. pdf

Schneider, S., *Science as a Contact Sport, Inside the Battle to Save Earth's Climate*, National Geographic Society, 2009

Schneider, S., Three Reports of the Intergovernmental Panel on Climate Change, *Environment*, 33 (1), 25, 1991

Siebenhüner, B., The Changing Role of Nation States in International Environmental Assessment – the Case of the IPCC, *Global Environmental Change* 13, 113, 2003

Sunstein, C.R., Montreal vs. Kyoto: A Tale of Two Procotols, 2006; http://papers.ssrn.com/ sol3/papers.cfm?abstract_id=913395

The Changing Atmosphere: Implications for Global Security – Conference Statement 1988; http://www.cmos.ca/ChangingAtmosphere1988e.pdf

Time, Science: One Big Greenhouse, May 28, 1956; http://www.time.com/time/magazine/ article/0,9171,937403,00.html

Trenberth, K.E., Stronger Evidence of Human Influence on Climate: The 2001 IPCC Assessment, *Environment* (May), 9, 2001

UK House of Lords, Select Committee on Economic Affairs, The Economics of Climate Change, July 6, 2005; http://www.publications.parliament.co.uk/pa/ld200506/ldselect/ ldconaf/12/12i.pdf

UNEP, *Declaration of the United Nations Conference on the Human Environment*; June 16, 1972; http://www.unep.org/Documents.Multilingual/Default.asp?documentid=9 7&articleid=1503

UNEP, *Multilateral Environmental Agreement Negotiator's Handbook*, 2007; http://unfccc. int/resource/docs/publications/negotiators_handbook.pdf

UNEP, Ozone Secretariat; http://www.unep.ch/ozone/index.shtml

UNFCCC, *Conference of the Parties, First Session – Part I*, Berlin, March 24, 1995; http:// unfccc.int/resource/docs/cop1/07.pdf

UNFCCC, *Conference of the Parties, First Session – Part II*, Berlin, March 28, 1995; http:// unfccc.int/resource/docs/cop1/07a01.pdf

UNFCCC, *Conference of the Parties, Second Session – Part I*, Geneva, October 29, 1996; http://unfccc.int/resource/docs/cop2/15.pdf

UNFCCC, *Conference of the Parties, Second Session – Part II*, Geneva, October 29,1996; http://unfccc.int/resource/docs/cop2/15a01.pdf

UNFCCC, *Conference of the Parties, Third Session – Part II*, Kyoto, March 25, 1998; http://unfccc.int/resource/docs/cop3/07a01.pdf

UNFCCC, *Convention Bodies, Subsidiary Bodies*; http://unfccc.int/essential_background/ convention/convention_bodies/items/2629.php

UNFCCC, *United Nations Framework Convention on Climate Change*, 1992; http://unfccc. int/resource/docs/convkp/conveng.pdf

United Nations Economic Commission for Europe, *Convention on Long-Range Transboundary Air Pollution*; http://www.unece.org/env/lrtap/

United Nations, *General Assembly, Forty-third Session*, Resolution 43/53, pp. 133–4, December 6, 1988; http://unfccc.int/resource/docs/1989/un/eng/ares4353.pdf; also see http://www.un.org/documents/ga/res/43/a43r053.htm

United Nations, *General Assembly, Forty-fourth Session, Protection of Global Climate For Present and Future Generations of Mankind*, September 19, 1989; http://unfccc.int/resource/docs/1989/un/eng/a44484.pdf

United Nations, General Assembly, *Forty-fourth Session*, Resolution 44/207, pp. 132–3, December 22, 1989; http://unfccc.int/resource/docs/1990/un/eng/ares44207.pdf

United Nations, *General Assembly, Forty-fifth Session, Protection of Global Climate For Present and Future Generations of Mankind*, November 2, 1990; http://unfccc.int/resource/docs/1990/un/eng/a45696.pdf

United Nations, *General Assembly, Forty-fifth Session*, Resolution 45/212, pp. 147–9, December 21, 1990; http://unfccc.int/resource/docs/1990/un/eng/ares45212.pdf

United Nations, *General Assembly, Forty-sixth Session*, Resolution 46/169, pp. 130–1, December 19, 1991; http://unfccc.int/resource/docs/1991/un/eng/ares46169.pdf

United Nations, *Intergovernmental Negotiating Committee for a Framework Convention on Climate Change, First Session*, New York, March 8, 1991; http://unfccc.int/resource/docs/a/06.pdf

United Nations, *Intergovernmental Negotiating Committee for a Framework Convention on Climate Change, Second Session*, August 19, 1991; http://unfccc.int/resource/docs/a/09.pdf

United Nations, *Intergovernmental Negotiating Committee for a Framework Convention on Climate Change, Third Session*, October 25,1991; http://unfccc.int/resource/docs/a/12.pdf

United Nations, *Intergovernmental Negotiating Committee for a Framework Convention on Climate Change, Third Session, Compilation of Proposals Related to Commitments*, August 15, 1991; http://unfccc.int/resource/docs/1991/a/eng/misc07.pdf

United Nations, *Intergovernmental Negotiating Committee for a Framework Convention on Climate Change, Fourth Session*, January 29, 1992; http://unfccc.int/resource/docs/a/15_1.pdf

United Nations, *Intergovernmental Negotiating Committee for a Framework Convention on Climate Change, Fourth Session, Joint Statement of the Group of 77*, December 19, 1991; http://unfccc.int/resource/docs/1991/a/eng/wg1l08.pdf

United Nations, *Intergovernmental Negotiating Committee for a Framework Convention on Climate Change, Fifth Session – Part I*, New York, March 10, 1992; http://unfccc.int/resource/docs/a/18p1.pdf

United Nations, *Intergovernmental Negotiating Committee for a Framework Convention on Climate Change, Fifth Session – Part II*, New York, May 15, 1992; http://unfccc.int/resource/docs/a/18p2.pdf

University of Toronto, Department of Economics, Unsung Canadian Achievement: the Late Professor John Dales, Inventor of Emissions Trading, Trade Offs, Fall 2007; http://northworthy.blogspot.com/2008/02/unsung-canadian-achievement-late.html

US House of Representatives, Hearing before the Subcommittee on Energy and Power, 105th Congress, Serial Nos. 105–108, *The Kyoto Protocol and Its Economic Implications*, March 4, 1998

Westrup, H., *Maurice Strong*, The Millbrook Press, 1994

WMO, *Resolutions of Congress and the Executive Council*, WMO-No. 508, 2012; ftp://ftp.wmo.int/Documents/MediaPublic/Publications/Policy_docs/WMO_508_en.pdf

Notes

1 UNEP, *Declaration of the United Nations Conference on the Human Environment*.
2 For a biography on Strong see Westrup, 1994.

3 Dowdeswell was Executive Director of UNEP from 1992–8, and was followed by Klaus Töpfer (1938–) (1998–2006); the current director is Achim Steiner (1961–).

4 For an excellent overview, see Bodansky, 1999; for a detailed report see UNEP, *Multilateral Environmental Agreement Negotiator's Handbook*, 2007.

5 The first CMP for the UNFCCC was held in 2005 in Montreal.

6 United Nations Economic Commission for Europe.

7 In June 2012, the UNEC presented an update of the LRTAP to the IPCC.

8 Ramus, 1991, pp. 15–16.

9 Ibid., 1991, pp. 1, 5.

10 Ibid., 1991, pp. 6, 26.

11 Ibid, 1991, p. 29.

12 Cass, 2006, p. 29.

13 Hilsenrath, August 13, 2009; the theory was later developed by David Montgomery.

14 University of Toronto, 2007.

15 Dales, 2002, p. 38.

16 McCauley *et al.*, 2008; Pooley, 2010, pp. 74–80.

17 Tg = teragrams, where "tera" comes from the Greek word for monster; tera = $1,000^4$; the SI unit is commonly used in this book; on more familiar terms, the unit of weight is the same as *million tonnes* (Mt).

18 EPA, *Clean Air Markets*.

19 US House of Representatives, Hearing before the Subcommittee on Energy and Power, 105th Congress, March 4, 1998. p. 75

20 Hilsenrath, August 13, 2009.

21 National Research Council, National Academy of Sciences, *Halocarbons: Effects of Stratospheric Ozone*.

22 UNEP, Ozone Secretariat.

23 The European Union did not actually form until November 1, 1993; from 1963 until this time, it was known as the European Community.

24 In an ironic twist, CFCs and their substitutes, such as hydrochlorofluorocarbons (HCFCs) and hydrofluorocarbons (HFCs), were discovered to be among the most significant contributors to climate change on a molecular basis, having thousands of times more impact on climate change than CO_2.

25 Borione and Ripert, in Mintzer and Leonard, 1994, p. 91; also see Sunstein, 2006.

26 The work was presented in 1895 and published the following year in *On the Influence of Carbonic Acid in the Air upon the Temperature on the Ground*; see Fleming, 1998, p. 76 (carbonic acid is carbon dioxide dissolved in water); the book by Fleming presents a detailed review of the early science of climate change and the greenhouse gas effect; Fleming had worked with Gilbert Plass.

27 Fleming, 1998, p. 74.

28 Ibid., p. 117.

29 For a review of the history, see ibid.; Bolin, 2007; Schneider, 2009.

30 Fleming, 1998, p. 121.

31 *Time*, May 28, 1956.

32 Hevesi, January 4, 2008.

33 See: http://www.wmo.int/pages/index_en.html

34 See: http://www.wmo.int/pages/prog/wcp/index_en.html

35 Davies, 1972, p. 332.

36 Wiin-Nielsen became an opponent of the IPCC; see Bolin, 2007, p. 136.

37 Obasi was succeeded by Michel Jarraud (1952–).

38 Jim Bruce of Canada was the Chairman of the conference; see Bruce 2001; Franz, 1997, presents a general overview of developments in this period with specific reference to Villach; also see European Commission, "The Greenhouse Effect and the Community," *Communication*, p. 56; follow-up meetings took place in Brussels, November 1986; Villach, October 1987; and Bellagio, November 1987.

39 Brundtland Commission's Report, 1987, Section 7.2, paragraph 23; Section 7.8, paragraph 116.
40 Bolin, 2007, p. 48.
41 *The Changing Atmosphere*, 1988, 292; also see Franz, 1997, p. 26.
42 Bolin, 2007, pp. 48–9; also see Schneider, 2009, p. 117.
43 Schneider, 2009, p. 117.
44 Cass, 2006, pp. 22, 25.
45 IPCC, November 9–11, 1988, Annex III, p. 6; for the role of Malta with the early years of the IPCC, see Bolin, 2007, p. 50.
46 United Nations, *General Assembly, Forty-third Session*, Resolution 43/53.
47 United Nations, *General Assembly, Forty-fourth Session*, Resolution 44/207.
48 United Nations, *General Assembly, Forty-fifth Session*, Resolution 45/212.
49 United Nations, *General Assembly, Forty-sixth Session*, Resolution 46/169; later Resolutions of this series were: 47/195 of December 22, 1992, 48/189 of December 21, 1993, 49/120 of December 19, 1994, 50/115 of December 20, 1995, 51/184 of December 16, 1996.
50 IPCC, *Report of the Third Session of WMO/UNEP Intergovernmental Panel on Climate Change*, Appendix D, Section 29.
51 IPCC, *Report of the Second Session of WMO/UNEP Intergovernmental Panel on Climate Change*, pp. 1, 3.
52 Agrawala, Structural and Process History of the Intergovernmental Panel on Climate Change, 1998, p. 634.
53 IPCC, *Report of the Fourth Session of WMO/UNEP Intergovernmental Panel on Climate Change*, p. 6.
54 IPCC, *First Assessment Report, Overview*, p. 60.
55 United Nations, *General Assembly, Forty-fifth Session*, Resolution 45/212.
56 United Nations, *General Assembly, Forty-fifth Session, Protection of Global Climate For Present and Future Generations of Mankind*.
57 United Nations, *Intergovernmental Negotiating Committee for a Framework Convention on Climate Change, First Session*, p. 24.
58 UNEP, *Declaration of the United Nations Conference on the Human Environment*.
59 Bodansky, in Mintzer and Leonard, 1994, p. 58.
60 For a general discussion, see Parks and Roberts, 2010.
61 United Nations, *Intergovernmental Negotiating Committee for a Framework Convention on Climate Change, First Session*, pp. 3–4.
62 Ibid., p. 13.
63 Ibid.; also known as "ecological debt," see Parks and Roberts, 2010, pp. 137, 144.
64 United Nations, *Intergovernmental Negotiating Committee for a Framework Convention on Climate Change, First Session*, pp. 12, 14.
65 United Nations, *Intergovernmental Negotiating Committee for a Framework Convention on Climate Change, Second Session*, p. 9; also see pp. 12–14.
66 Kjellen, in Mintzer and Leonard, 1994, p. 171.
67 United Nations, *Intergovernmental Negotiating Committee for a Framework Convention on Climate Change, Third Session*, pp. 18, 35–6.
68 United Nations, *Intergovernmental Negotiating Committee for a Framework Convention on Climate Change, Third Session, Compilation of Proposals Related to Commitments*, pp. 10, 19, 25.
69 Ibid., pp. 13, 34.
70 Ibid., p. 35.
71 Ibid., p. 42.
72 United Nations, *Intergovernmental Negotiating Committee for a Framework Convention on Climate Change, Fourth Session*, p. 31.
73 United Nations, *Intergovernmental Negotiating Committee for a Framework Convention on Climate Change, Fourth Session, Joint Statement of the Group of 77*.

74 United Nations, *Intergovernmental Negotiating Committee for a Framework Convention on Climate Change, Fourth Session*, p. 18ff.
75 Ibid., p. 27.
76 Kjellen, in Mintzer and Leonard, 1994, p. 158.
77 United Nations, *Intergovernmental Negotiating Committee for a Framework Convention on Climate Change, Fifth Session – Part I*, pp. 28, 38.
78 Djoghlaf, in Mintzer and Leonard, 1994, pp, 101–2.
79 See Borione and Ripert, in Mintzer and Leonard, 1994, p. 88.
80 Bodansky, in Mintzer and Leonard, 1994, p. 67.
81 United Nations, *Intergovernmental Negotiating Committee for a Framework Convention on Climate Change, Fifth Session – Part II*.
82 Dowdeswell and Kinley, in Mintzer and Leonard, 1994, p. 118; Dasgupta, in Mintzer and Leonard, 1994, pp. 142, 153; Kjellen, in Mintzer and Leonard, 1994, p. 162.
83 United Nations, *Intergovernmental Negotiating Committee for a Framework Convention on Climate Change, Fifth Session – Part II*, p. 8.
84 UNFCCC, *United Nations Framework Convention on Climate Change*, p. 1.
85 UNFCCC, *United Nations Framework Convention on Climate Change, Objective, Principles*, p. 4.
86 For a recent review of the American view on equity, see Bose, November 29, 2012.
87 Officially called the United Nations Conference on Environment and Development; the Secretary-General of the conference was Maurice Strong.
88 Borione and Ripert, in Mintzer and Leonard, 1994, pp. 90–1.
89 Excellent summaries of the meetings since 1995 can be found in *Earth Negotiations Bulletin*; Issue 12 (21), April 10, 1995, reviews the daily events at Berlin.
90 UNFCCC, *Conference of the Parties, First Session – Part II*, 2 (b), p. 5; see Part I, paragraph 55, p. 23.
91 Annex I was a list of the developed countries that had emission targets.
92 UNFCCC, *Conference of the Parties, Second Session – Part I*, p. 5; also see *Part II*, p. 42.
93 For details of the Kyoto Protocol see Cole, 2008.
94 UNFCCC, *Conference of the Parties, Third Session – Part II*, paragraph 10, p. 16.
95 Ibid., paragraph 7, p. 11.
96 US House of Representatives, March 4, 1998, p. 15.
97 IPCC, *Report of the Fifth Session of WMO/UNEP Intergovernmental Panel on Climate Change*, pp. 2, 4, 9.
98 Ibid., p. 6.
99 The coordinator of the Second World Climate Conference was Howard L. Ferguson of Canada, who had been a major factor in the Toronto Conference in 1988, and was the Canadian Principal Delegate to the IPCC.
100 IPCC, *Report of the Sixth Session of the Intergovernmental Panel on Climate Change*, p. 7.
101 Ibid., p. 8.
102 Agrawala, Context and Early Origins, 1998, p. 611; Agrawala, Structural and Process History, 1998, p. 635.
103 IPCC, *Report of the 29th Session of the IPCC*, p. 22.
104 UNFCCC, *United Nations Framework Convention on Climate Change*, p. 20.
105 IPCC, *Second Assessment Report, Working Group I: The Science of Climate Change*, p. xi.
106 UNFCCC, *United Nations Framework Convention on Climate Change*, p. 4.
107 IPCC, *Report of the 25th Session of the IPCC, Review of IPCC Terms of Reference*, p. 1.
108 IPCC, *Second Assessment Report, Working Group I: Scientific Assessment of Climate Change*, p. xi.
109 IPCC, *Report of the Ninth Session of the Intergovernmental Panel on Climate Change*, Appendix F.

110 IPCC, *Report of the Ninth Session of the Intergovernmental Panel on Climate Change*, pp. 7–11.
111 Ibid., p. 3.
112 Ibid., p. 14.
113 Agrawala, Structural and Process History, 1998, p. 637.
114 Bolin, 2007, pp. 85–7.
115 Agrawala, Structural and Process History, 1998, p. 636.
116 Pearce, 1994.
117 IPCC, *Report of the Tenth Session of the Intergovernmental Panel on Climate Change*, p. 3.
118 IPCC, *Report of the Eighth Session of the Intergovernmental Panel on Climate Change*, pp. 4–6.
119 UNFCCC, *Convention Bodies, Subsidiary Bodies*.
120 IPCC, *Report of the Eleventh Session of the Intergovernmental Panel on Climate Change*, p. 3.
121 Ibid., p. 4.
122 *Earth Negotiations Bulletin*, 12 (248), September 13, 2004, pp. 3–4.
123 *Earth Negotiations Bulletin*, 12 (547), June 12, 2012, p. 6.
124 InterAcademy Council, 2010, p. vii.
125 Pearce, 2005, p. 52.
126 Hecht and Tirpak, 1995, pp. 380–1, 400; the actions of the Reagan Administration are discussed in the section "'Acting' on Climate Change," in the chapter "The Players."
127 IPCC, *Report of the 35th Session of the IPCC, Review of the IPCC Processes and Procedures, Remaining Issues Related to Governance and Management, Comments by the UNEP Secretariat and WMO on the Draft Terms of Reference of the IPCC Secretariat and the IPCC Technical Support Units Contained in Document IPCC-XXXV/Doc.12*, p. 1.
128 Agrawala, Context and Early Origins, 1998, p. 611.
129 Ibid., p. 615.
130 Ibid.; also see IPCC, *Report of the First Session of WMO/UNEP Intergovernmental Panel on Climate Change*, p. 6.
131 WMO, 2012, pp. II–28.
132 *The Changing Atmosphere*, 1988, p. 292; the IPCC was also discussed in the summer at a conference held at the University Corporation for Atmospheric Research, see Bolin, 2007, pp. 47–8.
133 United Nations, *General Assembly, Forty-Third Session*, Resolution 43/53; United Nations, *General Assembly, Forty-fourth Session, Protection of Global Climate For Present and Future Generations of Mankind*, pp. 1–2.
134 IPCC, *Report of the 35th Session of the IPCC, Review of the IPCC Processes and Procedures, Remaining Issues Related to Governance and Management, Memorandum of Understanding between UNEP and WMO on the IPCC*.
135 IPCC, *Report of the 35th Session of the IPCC, Review of the IPCC Processes and Procedures, Remaining Issues Related to Governance and Management, Comments by the UNEP Secretariat and WMO on the Draft Terms of Reference of the IPCC Secretariat and the IPCC Technical Support Units Contained in Document IPCC-XXXV/Doc.12*, p. 1; also see IPCC, *First Assessment Report, Overview*, p. 60.
136 InterAcademy Council, 2010, p. 44.
137 *Earth Negotiations Bulletin*, 12 (165), April 9, 2001, p. 5.
138 Agrawala, Structural and Process History, 1998, p. 622; Bolin, 2007, pp. 101, 107.
139 IPCC, *Report of the 25th Session of the IPCC, Review of IPCC Terms of Reference*; IPCC, *Report of the 26th Session of the IPCC, Review of IPCC Terms of Reference*; IPCC, *Report of the 26th Session of the IPCC, Review of IPCC Terms of Reference, Compilation of Comments Received on the Draft Proposal*; IPCC, *Report of the 26th*

Session of the IPCC, Review of IPCC Terms of Reference, Summary of Relevant Resolutions and Decisions.

140 Bolin, 2007, pp. 77.

141 IPCC, *Report of the Twentieth Session of the Intergovernmental Panel on Climate Change*, pp. 11–17.

142 IPCC, *Report of the Eleventh Session of the Intergovernmental Panel on Climate Change*, p. 5.

143 IPCC, *Report of the Thirteenth Session of the Intergovernmental Panel on Climate Change*, pp. 3–4; IPCC, *Report of the Twelfth Session of the Intergovernmental Panel on Climate Change*, pp. 5, 6; Bolin, 2007, p. 138.

144 IPCC, *Report of the 27th Session of the IPCC Bureau*, p. 4.

145 Leemans, 2008, p. 13; Siebenhüner, 2003, pp. 114–15.

146 IPCC, *Principles and Procedures of the IPCC*, Appendix A; also see Netherlands Environmental Assessment Agency, 2010, p. 37.

147 IPCC, *Report of the Second Session of WMO/UNEP Intergovernmental Panel on Climate Change*, p. 4; also see IPCC, *Report of the Fourteenth Session of the Intergovernmental Panel on Climate Change*, Appendix C; see Bolin, 2007, pp. 122–4.

148 IPCC, *First Assessment Report, Working Group I: Scientific Assessment of Climate Change*, p. v.

149 Article 2 later became a major issue of discussion; see IPCC, *Report of the 32nd Session of the IPCC, The IPCC Fifth Assessment Report (AR5), Progress Report of the Cross Working Group Consultation on Article 2 of the UNFCCC.*

150 IPCC, *Report of the Tenth Session of the Intergovernmental Panel on Climate Change*, p. 8.

151 *Earth Negotiations Bulletin*, 12 (177), October 2, 2001, p. 9.

152 IPCC, *Report of the Twentieth Session of the Intergovernmental Panel on Climate Change, Preparations for the Fourth Assessment Report*, p. 4.

153 Ibid., pp. 1, 4; IPCC, *Report of the 21st Session of the IPCC, A Proposal for the Synthesis Report (SYR) for the AR4*, p. 3.

154 IPCC, *Report of the First Session of WMO/UNEP Intergovernmental Panel on Climate Change*, pp. 2, 4; also see United Nations, *General Assembly, Forty-fourth Session, Protection of Global Climate For Present and Future Generations of Mankind*, pp. 5, 6.

155 United Nations, *General Assembly, Forty-fourth Session, Protection of Global Climate For Present and Future Generations of Mankind*, p. 6; IPCC, *Report of the First Session of WMO/UNEP Intergovernmental Panel on Climate Change*, Annex IV.

156 IPCC, *Report of the Seventeenth Session of the Intergovernmental Panel on Climate Change, Future Work Programme of the IPCC*, p. 3.

157 IPCC, *Request for AR4 Nominations*, Attachment 2; IPCC, *Report of the Twentieth Session of the Intergovernmental Panel on Climate Change*, p. 14, Annex 1.

158 IPCC, *Report of the Ninth Session of the Intergovernmental Panel on Climate Change*, Appendix G.

159 IPCC, *Report of the Tenth Session of the Intergovernmental Panel on Climate Change*, Appendix C.

160 IPCC, *Report of the Ninth Session of the Intergovernmental Panel on Climate Change*, Appendix G.

161 IPCC, *Report of the Second Session of WMO/UNEP Intergovernmental Panel on Climate Change*, p. 11

162 Ibid., p. 10.

163 IPCC, *Report of the Tenth Session of the Intergovernmental Panel on Climate Change*, p. 7; a designated government contact from each member country to be the point of contact for all correspondence with the IPCC and its Secretariat.

164 IPCC, *Request for AR4 Nominations, Letter to Governments Requesting Information.*

165 IPCC, *Report of the Fifth Session of WMO/UNEP Intergovernmental Panel on Climate Change*, p. 8.
166 IPCC, *Report of the Nineteenth Session of the Intergovernmental Panel on Climate Change, IPCC Programme and Budget for 2003–2005*, p. 2; IPCC, *Report of the Nineteenth Session of the Intergovernmental Panel on Climate Change, Evaluation of the Work of Working Group III and Its Technical Support Unit from 1998–2002: Lessons for the Future*, p. 1; IPCC, *Report of the Twentieth Session of the Intergovernmental Panel on Climate Change, Preparations for the Fourth Assessment Report*, pp. 4–5.
167 IPCC, *Report of the 33rd Session of the IPCC, Decisions Taken with Respect to the Review of IPCC Processes and Procedures, Procedures*, p. 1.
168 An excellent review of the peer-review process appears in Russell *et al.*, July 2010, (p. 64, Appendix 5).
169 IPCC, *Report of the Tenth Session of the Intergovernmental Panel on Climate Change*, p. 9.
170 Agrawala, Structural and Process History, 1998, pp. 623–4.
171 InterAcademy Council, 2010, p. 3.
172 Pachauri, 2010.
173 IPCC, *Report of the First Session of WMO/UNEP Intergovernmental Panel on Climate Change*, p. 1; Jim Bruce later became co-chair of WG III, where he replaced Elizabeth Dowdeswell after she became Executive Director of UNEP.
174 IPCC, *Report of the First Session of WMO/UNEP Intergovernmental Panel on Climate Change*, Annex I.
175 IPCC, *Principles and Procedures of the IPCC*, Appendix A; also see IPCC, *Report of the Fourteenth Session of the Intergovernmental Panel on Climate Change*, Annex C.
176 IPCC, *Report of the Ninth Session of the Intergovernmental Panel on Climate Change*, Appendix G; see, for example, IPCC, *Report of the Tenth Session of the Intergovernmental Panel on Climate Change*, pp. 5–6; IPCC, *Report of the Eleventh Session of the Intergovernmental Panel on Climate Change*, p. 2.
177 Jung, 1999, p. 20.
178 Agrawala, Structural and Process History, 1998, p. 627.
179 Hecht and Tirpak, 1995, p. 387.
180 IPCC, *Third Assessment Report, Working Group I: The Scientific Basis*, p. ix.
181 Trenberth, 2001, p. 11.
182 IPCC, *Report of the Eighth Session of the Intergovernmental Panel on Climate Change*, p. 16; IPCC, *Report of the 33rd Session of the IPCC, Decisions Taken with Respect to the Review of IPCC Processes and Procedures, Governance and Management*, p. 5; a draft of Terms of Reference for the Bureau was issued, IPCC, *Report of the 33rd Session of the IPCC, Review of the IPCC Processes and Procedures, Draft Proposal by the Task Group on Governance and Management, Terms of Reference for the Bureau and Bureau Members*.
183 IPCC, *Report of the First Session of WMO/UNEP Intergovernmental Panel on Climate Change*, pp. 5–6.
184 IPCC, *Report of the Eighth Session of the Intergovernmental Panel on Climate Change*, p. 11.
185 IPCC, *Report of the 28th Session of the IPCC*.
186 IPCC, *Report of the First Session of WMO/UNEP Intergovernmental Panel on Climate Change*, p. 9; United Nations, *General Assembly, Forty-fourth Session, Protection of Global Climate For Present and Future Generations of Mankind*, p. 5; IPCC, *First Assessment Report, Working Group I: Scientific Assessment of Climate Change*, Appendix 2.
187 IPCC, *Report of the Twelfth Session of the Intergovernmental Panel on Climate Change*, p. 16.

188 Bolin, 2007, p. 145.
189 IPCC, *Report of the Twelfth Session of the Intergovernmental Panel on Climate Change*, Appendix K.
190 IPCC, *Report of the Tenth Session of the Intergovernmental Panel on Climate Change*, Appendix I.
191 IPCC, *Report of the Seventeenth Session of the Intergovernmental Panel on Climate Change, Future Work Programme of the IPCC*, pp. 3–4.
192 IPCC, *Report of the Eighteenth Session of the Intergovernmental Panel on Climate Change, Future of the Intergovernmental Panel on Climate Change*, p. 2.
193 IPCC, *Report of the Thirteenth Session of the Intergovernmental Panel on Climate Change*, pp. 8–9.
194 IPCC, *Report of the Nineteenth Session of the Intergovernmental Panel on Climate Change*, Appendix A.
195 IPCC, *Report of the Nineteenth Session of the Intergovernmental Panel on Climate Change*, pp. 2–3.
196 Bolin, 2002, p. 1235.
197 See, for example, Bolin, 2007, p. 186.
198 http://www.teriin.org/index.php
199 For example, see Bolin, 2007, pp. 49–50.
200 IPCC, *Report of the Thirteenth Session of the Intergovernmental Panel on Climate Change*, p. 8.
201 IPCC, *Report of the 22nd Session of the IPCC, Rules and Procedures for the Election of the IPCC Bureau and Task Force Bureau*, p. 17.
202 IPCC, *Report of the 24th Session of the IPCC*, Annex 4.
203 IPCC, *Report of the 24th Session of the IPCC, Revised Draft Rules and Procedures (ROP) for the Election of the IPCC Bureau and Task Force Bureau*, Annex B.
204 *Earth Negotiations Bulletin*, 12 (278), September 30, 2005, p. 8.
205 IPCC, *Report of the 31st Session of the IPCC, Revision of the Rules of Procedures for the Election of the IPCC Bureau and Any Task Force Bureau, Task Group on Rules of Procedures for the Election of the IPCC Bureau and Any Task Force Bureau*, p. 2; IPCC, *Report of the 32nd Session of the IPCC, Rules of Procedures for the Election of the IPCC Bureau and Any Task Force Bureau*, p. 61.
206 IPCC, *Report of the 34th Session of the IPCC, Further Work to Adopting Revisions to "Appendix C of the Principles Governing IPCC Work: Rules of Procedures for the Election of the IPCC Bureau and Any Task Force Bureau"*, pp. 15–16; IPCC, *Report of the 35th Session of the IPCC, Procedures*, Appendix C.
207 IPCC, *Report of the 35th Session of the IPCC, Procedures*, Appendix C, p. 5.
208 IPCC, *Report of the Eighth Session of the Intergovernmental Panel on Climate Change*, pp. 16–17.
209 IPCC, *Report of the 21st Session of the IPCC, Election Procedures for the IPCC Bureau and TFB*.
210 IPCC, *Report of the 22nd Session of the IPCC, Rules and Procedures for the Election of the IPCC Bureau and Task Force Bureau*; IPCC, *Report of the 24th Session of the IPCC, Revised Draft Rules and Procedures (ROP) for the Election of the IPCC Bureau and Task Force Bureau*; IPCC, *Report of the 24th Session of the IPCC*, Annex 4; IPCC, *Report of the 25th Session of the IPCC, Draft Rules of Procedures for the Election of the IPCC Bureau and Any Task Force Bureau*.
211 IPCC, *Report of the 25th Session of the IPCC*, pp. 2–3, Annex 5.
212 IPCC, *Report of the 29th Session of the IPCC*, p. 3.
213 *Earth Negotiations Bulletin*, 12 (384), September 6, 2008, p. 1.
214 IPCC, *Report of the 30th Session of the IPCC, Revision of the Rules of Procedures for the Election of the IPCC Bureau and Any Task Force Bureau*; IPCC, *Report of the 31st Session of the IPCC, Revision of the Rules of Procedures for the Election of the IPCC*

Bureau and Any Task Force Bureau; IPCC, *Report of the 32nd Session of the IPCC, Rules of Procedures for the Election of the IPCC Bureau and Any Task Force Bureau.*

215 InterAcademy Council, 2010, p. 46.

216 IPCC, *Report of the 33rd Session of the IPCC, Decisions Taken with Respect to the Review of IPCC Processes and Procedures, Governance and Management*, p. 3; IPCC, *Report of the 34th Session of the IPCC, Further Work to Adopting Revisions to "Appendix C of the Principles Governing IPCC Work: Rules of Procedures for the Election of the IPCC Bureau and Any Task Force Bureau"*, p. 1.

217 Edwards and Schneider, 1997, p. 7.

218 Ibid.

219 IPCC, *Report of the Fifth Session of WMO/UNEP Intergovernmental Panel on Climate Change*, p. 8.

220 Bolin, 2007, p. 70.

221 IPCC, *Report of the Ninth Session of the Intergovernmental Panel on Climate Change*, Appendix G.

222 IPCC, *Report of the Fourteenth Session of the Intergovernmental Panel on Climate Change*, Appendix F; for minor amendments see IPCC, *Report of the 21st Session of the IPCC*, Annex 7, and IPCC, *Report of the 21st Session of the IPCC, Review of IPCC Principles.*

223 IPCC, *Report of the 21st Session of the Intergovernmental Panel on Climate Change*, pp. 7, 12–17.

224 IPCC, *Report of the Twentieth Session of the Intergovernmental Panel on Climate Change*, p. 7, Annex 7; IPCC, *Report of the 21st Session of the IPCC, Review of IPCC Principles*; IPCC, *Report of the 22nd Session of the IPCC*, p. 4.

225 InterAcademy Council, 2010, p. xvi.

226 IPCC, *Report of the 28th Session of the IPCC*, p. 5.

227 IPCC, *Report of the 28th Session of the IPCC, Future of the IPCC, Compiled Comments from Governments, Authors, Organizations and Bureau Members*; IPCC, *Report of the 28th Session of the IPCC, Future of the IPCC, Comments Received after the Deadline and Translations into English of Earlier Submissions*; IPCC, *Report of the 28th Session of the IPCC, Future of the IPCC, Synthesis of Comments.*

228 IPCC, *Report of the 30th Session of the IPCC*, pp. 1–4.

229 IPCC, *Draft Report of the 32nd Session of the IPCC*, pp. 9–15; Annex 3, pp. 63–4.

230 IPCC, *Report of the 33rd Session of the IPCC, Review of the IPCC Processes and Procedures, Proposal by the Task Group on Procedures.*

231 IPCC, *Report of the 33rd Session of the IPCC, Decisions Taken with Respect to the Review of IPCC Processes and Procedures, Procedures*; IPCC, *Report of the 34th Session of the IPCC, Review of IPCC Processes and Procedures, Procedures: Adoption of the revised "Appendix A to the Principles Governing IPCC Work: Procedures for the Preparation, Review, Acceptance, Adoption, Approval and Publications of IPCC Reports"*; IPCC, *Report of the 35th Session of the IPCC, Procedures*, Appendix A.

232 Agrawala, Context and Early Origins, 1998, p. 611.

233 IPCC, *Report of the Sixth Session of the Intergovernmental Panel on Climate Change*, p. 14.

234 IPCC, *Report of the Eighth Session of the Intergovernmental Panel on Climate Change*, p. 5.

235 Bolin, 2007, p. 83.

236 IPCC, *Report of the Ninth Session of the Intergovernmental Panel on Climate Change*, p. 4.

237 See, for example, IPCC, *Report of the Third Session of WMO/UNEP Intergovernmental Panel on Climate Change*, p. 5 for comments by Obasi, and pp. 8–9 for comments by Tolba. Tolba was especially active – see IPCC, *Report of*

the Fourth Session of WMO/UNEP Intergovernmental Panel on Climate Change, p. 7; IPCC, *Report of the Fifth Session of WMO/UNEP Intergovernmental Panel on Climate Change*, p. 5.

238 IPCC, *Report of the First Session of WMO/UNEP Intergovernmental Panel on Climate Change*, p. 2.

239 IPCC, *Report of the Fourth Session of WMO/UNEP Intergovernmental Panel on Climate Change*, p. 12.

240 IPCC, *Principles and Procedures of the IPCC*.

241 Bolin, 1994, p. 27.

242 Ibid.

243 IPCC, *Report of the Third Session of WMO/UNEP Intergovernmental Panel on Climate Change*, p. 9.

244 IPCC, *Report of the Tenth Session of the Intergovernmental Panel on Climate Change*, p. 3.

245 *Earth Negotiations Bulletin*, 12 (363), April 13, 2008, p. 8.

246 The UK House of Lords was critical of this process; see UK House of Lords, Select Committee on Economic Affairs, *The Economics of Climate Change*, July 6, 2005, p. 59.

247 IPCC, *Report of the Sixteenth Session of the Intergovernmental Panel on Climate Change*, Annex C.

248 IPCC, *Report of the Seventeenth Session of the Intergovernmental Panel on Climate Change, Future Work Programme of the IPCC*, p. 3.

249 *Earth Negotiations Bulletin*, 12 (342), November 19, 2007, p. 14.

250 *Earth Negotiations Bulletin*, 12 (177), October 2, 2001, p. 3; for another review of the TAR see Schneider, 2009, pp. 165–72.

251 IPCC, *Report of the 27th Session of the IPCC, IPCC Fourth Assessment Report "Climate Change 2007" Draft Synthesis Report*; for a discussion of the AR4 meeting in Brussels see Schneider, 2009, pp. 180–97.

252 *Earth Negotiations Bulletin*, 12 (342), November 19, 2007, pp. 3–12.

253 Ibid., p. 10.

254 Ravindranath, 2010, p. 34; also see p. 30; for example, see Schneider, 2009, pp. 183–5.

255 InterAcademy Council, 2010, p. 24; IPCC, *Report of the 28th Session of the IPCC, Future of the IPCC, Synthesis of Comments*, p. 2.

256 IPCC, *Report of the First Session of WMO/UNEP Intergovernmental Panel on Climate Change*, p. 6.

257 IPCC, *Report of the First Session of WMO/UNEP Intergovernmental Panel on Climate Change*, p. 8.

258 Schneider, 1991, p. 25; also see Schneider, 2009, pp. 124–5.

259 IPCC, *Report of the First Session of WMO/UNEP Intergovernmental Panel on Climate Change*, p. 5; Agrawala, Structural and Process History, 1999, p. 628.

260 IPCC, *Report of the Second Session of WMO/UNEP Intergovernmental Panel on Climate Change*, p. 22.

261 IPCC, *Report of the Third Session of WMO/UNEP Intergovernmental Panel on Climate Change*, Appendix F.

262 IPCC, *Report of the Second Session of WMO/UNEP Intergovernmental Panel on Climate Change*, pp. 2, 3; also see p. 5 and Annex III, p. 1.

263 IPCC, *Report of the Second Session of WMO/UNEP Intergovernmental Panel on Climate Change*, Annex III, p. 1; also see Annex IV, p. 1.

264 Ibid., Annex III, pp. 18–20.

265 Ibid., pp. 19–20.

266 United Nations, *General Assembly*, Forty-Fourth Session, Resolution 44/207, p. 131.

267 IPCC, *Report of the Fourth Session of WMO/UNEP Intergovernmental Panel on Climate Change*, p. 13; also see IPCC, *Report of the Third Session of WMO/UNEP*

Intergovernmental Panel on Climate Change, pp. 25–6; Appendix E; United Nations, *General Assembly, Forty-fourth Session, Protection of Global Climate For Present and Future Generations of Mankind*, p. 6.

268 IPCC, *Report of the Fourth Session of WMO/UNEP Intergovernmental Panel on Climate Change*, p. 6.

269 Ibid., p. 5

270 Ibid.

271 IPCC, *First Assessment Report, Overview*, pp. 59–60.

272 IPCC, *First Assessment Report, Working Group II: Impacts Assessment of Climate Change, Forward*.

273 Agrawala, Structural and Process History, 1998, p. 634; also see Bolin, 2007, p. 68.

274 IPCC, *Report of the Fifth Session of WMO/UNEP Intergovernmental Panel on Climate Change*, p. 5.

275 Ibid.; also see pp. 2–3.

276 Ibid., p. 8.

277 IPCC, *Report of the Sixth Session of the Intergovernmental Panel on Climate Change*, p. 24.

278 IPCC, *Report of the Fifth Session of WMO/UNEP Intergovernmental Panel on Climate Change*, p. 6

279 IPCC, *Report of the Sixth Session of the Intergovernmental Panel on Climate Change*, p. 16.

280 IPCC, *Report of the Tenth Session of the Intergovernmental Panel on Climate Change, Appendix, Long-Term IPCC Funding*.

281 IPCC, *Report of the Nineteenth Session of the Intergovernmental Panel on Climate Change, Options for the IPCC Budget Management*, p. 1.

282 IPCC, *Report of the Sixth Session of the Intergovernmental Panel on Climate Change*, p. 4.

283 Ibid., pp. 4–5.

284 Ibid., p. 5.

285 Ibid., p. 8.

286 Ibid., p. 9.

287 Ibid., p. 10; see IPCC, *Report of the Eighth Session of the Intergovernmental Panel on Climate Change*, pp. 11–16, Appendix E; Bolin, 2007, p. 146.

288 IPCC, *Report of the Eighth Session of the Intergovernmental Panel on Climate Change*, p. 11, Appendix E.

289 Ibid., pp. 11–16, Appendix E.

290 IPCC, *Report of the Ninth Session of the Intergovernmental Panel on Climate Change*, p. 3, Appendix G.

291 Ibid., p. 1.

292 IPCC, *Report of the Tenth Session of the Intergovernmental Panel on Climate Change*, pp. 1, 9.

293 IPCC, *Report of the Thirteenth Session of the Intergovernmental Panel on Climate Change*, pp. 8–9, Appendix B.

294 *Earth Negotiations Bulletin*, 12 (177), October 2, 2001, p. 8.

295 An exception was the Eighteenth Session, where several delegates wanted a new group to study the participation of developing countries; *Earth Negotiations Bulletin*, 12 (177), October 2, 2001, p. 8.

296 IPCC, *Report of the 30th Session of the IPCC*, p. 3.

297 IPCC, *Report of the 31st Session of the IPCC, Improving Participation of Developing/ EIT Countries in the IPCC*, pp. 2–3.

298 Karlsson *et al.*, 2007, and references therein.

299 See Hulme and Mahony, 2010, p. 709 and references therein.

300 InterAcademy Council, 2010, p. 67; also see Hulme and Mahony, 2010, pp. 708–9.

301 IPCC, *Report of the 31st Session of the IPCC, Implementation of Decisions Taken by IPCC-30, Involving Developing/EIT Country Scientists*, p. 3.

302 Bolin, 2007, p. 84.

303 IPCC, *Report of the Nineteenth Session of the Intergovernmental Panel on Climate Change, Evaluation of the Work of Working Group III and Its Technical Support Unit from 1998–2002: Lessons for the Future*, p. 2.

304 Price, 2010, p. 732; also see Pearce, June 26, 2012; and IPCC, *IPCC Statement on New Scientist Article*, Statements, June 24, 2012, Statements.

305 IPCC, *Report of the Sixth Session of the Intergovernmental Panel on Climate Change*, p. 9.

306 All web-based references were confirmed (accessed) in June 2013, unless otherwise stated.

307 IPCC *Assessment Reports* can be found at: http://www.ipcc.ch/publications_and_data/publications_and_data_reports.shtml.

308 IPCC Plenary Session minutes and supporting documents can be found at: http://www.ipcc.ch/meeting_documentation/meeting_documentation_ipcc_sessions_and_ipcc_wgs_sessions.shtml.

309 IPCC Bureau Session minutes can be found at: http://www.ipcc.ch/meeting_documentation/meeting_documentation_sessions_of_the_ipcc_bureau.shtml.

310 For the report of the 32nd Session and documents for the 33rd Sessions see: http://www.ipcc.ch/scripts/_session_template.php?page=_33ipcc.htm.

311 IPCC Statements can be found at: http://www.ipcc.ch/news_and_events/press_information.shtml.

312 InterAcademy Council, Renate, C., *IPCC Products, Procedures, and Processes, Secretary of the IPCC Renate Christ Presentation to the IAC Review Committee*, May 10, 2010, slide 17.

2 Foul ball

The 2010 season

The driver for climate change legislation has been (largely) the information contained in the Assessment Reports of the IPCC. This dependence places incredible importance on the accuracy of the information. However, there was a problem: the issue was not *if* a mistake would be made, but *when*. There were four fundamental weaknesses within the IPCC:

1. scope of the literature assessment and review
2. time pressure from the climate negotiations
3. dilution of scientific expertise by forced regional representation
4. lack of formal qualifications and standard operating procedures within the IPCC.

Errors were bound to happen under such circumstances, which could undermine the credibility of the IPCC and derail climate change legislation. Compounding the risk, there were well-funded climate skeptics who were eager to exploit any flaws within the system.

"Eightgate"

One word in the SAR of WG I, "discernible" (with regard to the human influence on climate change),[1] was arguably the most important aspect of the entire document. This important development allowed for a separation between natural and anthropogenic climate change:[2]

> Our ability to quantify the human influence on global climate is currently limited because the expected signal is still emerging from the noise of natural variability, and because there are uncertainties in key factors. These include the magnitude and patterns of long term natural variability and the time-evolving pattern of forcing by, and response to, changes in concentrations of greenhouse gases and aerosols, and land surface changes. Nevertheless, the balance of evidence suggests that there is a discernible human influence on global climate.

The wording had originated from Chapter 8, *Detection of Climate Change and Attribution of Causes*, where Benjamin D. Santer (1955–) had been the CLA.

This highlight of the SAR almost didn't make it! A lot of the discussion centered on Chapter 8 during the Plenary Session of WG I in Madrid. The delegate from Kenya suggested that the entire chapter be removed.[3] Saudi Arabia, with help from the climate skeptic Don Pearlman of the World Climate Council, had also argued against it.[4] The conversation became heated as Mohammed Al-Saban, the Saudi delegate, told Santer: "I am a representative of a sovereign nation and you are just a scientist."[5]

The Kenyan delegate, who had previously called for the removal of all of Chapter 8, now supported Santer, defusing the situation, and the wording remained.

The following appeared in the minutes of the meeting of the Eleventh Session of the IPCC when the SAR was accepted (December 1995):[6]

> Sir John Houghton informed the Panel that the draft SPM submitted to the Working Group at its fifth session [in Madrid] had contained a draft Executive Summary. The latter was expanded by the Working Group in the approval process into a new SPM and the material in the (old) draft SPM that was not included in the approved SPM was compiled into a Technical Summary of the Report of the Working Group by the lead authors.

The wording is unclear, but the Chairman of WG I was likely informing delegates that changes were made to the SPM after the Madrid acceptance meeting.

The SAR was published in early 1996. A coordinated effort by major skeptic groups was launched against Santer, and Chapter 8, known as "eightgate," became the focus of one of the early inflammatory attacks against the IPCC and its contributors. In early June 1996, the Global Climate Coalition (GCC) raised a "big stink"[7] about the SAR. The association of major corporations attacked the integrity of the Assessment Report by charging the IPCC with "scientific cleansing," as Chapter 8 had been amended after it had been approved by the Plenary Session of WG I. The complaint had not just gone to Bolin but also to members of Congress. This was followed on June 12 by accusations from the Chairman and founder of the Marshall Institute, Fredrick Seitz (1911–2008), a noted scientist and climate change skeptic, in the *Wall Street Journal*:[8]

> But this report is not what it appears to be – it is not the version that was approved by the contributing scientists listed on the title page. In my more than 60 years as a member of the American scientific community, including service as president of both the National Academy of Sciences and the American Physical Society, I have never witnessed a more disturbing corruption of the peer-review process than the events that led to this IPCC report.

Santer and other scientific contributors defended the SAR,[9] as did Bolin.[10] The debate, though, played outside of scientific circles in the public press,[11] an arena that fell within the strengths of lobbying groups such as the Marshall Institute. The CLA of Chapter 8 and the other scientists found themselves in unfamiliar territory:[12]

> Santer, who felt forced to defend himself, spent the majority of his summer time responding to the charges. Previously a quiet, private man known to scientists primarily as a proponent of the rigorous use of statistical methods, Santer rapidly became a public figure, submitting to dozens of interviews. The drain on his time and energy during this period kept him from his scientific work, he said.

Santer was clearly upset by the allegations: "I am really troubled by what is going on. This appears to be a skillful campaign to discredit the IPCC, me, and my reputation as a scientist."[13] Eightgate was a wake-up call, warning the scientists that the carbon game could be played dirty; one where empirical research and knowledge building would not necessarily win the day. The skeptics' war on the IPCC was just getting started. It would haunt future Assessment Reports and Lead Authors.

The accusatory tone, personal attacks and implied intent of the criticisms against Santer were unfair, but Seitz had a point: Santer had changed the main text of the report after it had been "accepted" at the Plenary Session of WG I. An article in the prestigious journal *Nature* supported the IPCC and Santer, and discounted the GCC comments, but suggested that there was cause for concern:[14]

> The complaints are not entirely groundless. IPCC officials claim that the sole reason for the revisions was to tidy up the text, and in particular to ensure that it conformed to a "policymakers summary" of the full report that was tortuously agreed by government delegates at the Madrid meeting. But there is some evidence that the revision process did result in a subtle shift in the relative weight given to different types of arguments, and that – not surprisingly – this shift tended to favour arguments that aligned with the report's broad conclusions. Conversely, some phrases that might have been (mis)interpreted as undermining these conclusions, particularly if, as IPCC officials feared, they were taken out of context, have disappeared.

An analysis of the scandal was performed by Stephen Schneider, who had been a LA in the SAR of WG I, and Paul Edwards. They concluded that there had been a problem: it was not the actions of Santer but the rules of the IPCC:[15]

> Santer *et al.* believed that they were following IPCC rules, and this made perfect sense within the well-established informal culture of the IPCC. However, a careful reading of the IPCC's formal rules reveals that in fact the rules neither allow nor prohibit changes to a report after its formal acceptance. The legalistic Seitz/GCC reading of the rules is not, therefore, completely implausible – even if it was, as we believe, primarily a smokescreen to divert

attention from the clear consensus that attribution could no longer be considered unlikely.

Our analysis suggests a significant flaw in the rules as currently written. While "approved" documents (the SPMs) clearly must not be altered once approved, there is no precisely defined closure mechanism for "accepted" documents (full length Working Group reports and their constituent chapters). The Seitz/GCC attack has effectively demonstrated that a hybrid science/policy organization like the IPCC needs better, more explicit rules of procedure. This minor virtue aside, however, the Seitz/GCC reading violates the spirit and intent of the IPCC process.

The following year a major change was made to the Assessment Report process.[16] In the *Decision Paper* for the TAR by Watson, the new position of Review Editor was created, with these acting as advisors to the Lead Authors on "controversial issues."[17]

Some issues also arose in the preparation of the TAR. In a candid review from 2001 and 2002, the Chairmen of WG III complained that controversies tended to dominate the entire process, which distracted from other important issues.[18] Especially disconcerting was criticism of the writing teams, including:[19]

- capabilities of CLAs to coordinate team and to work towards consensus documents not well enough known in advance
- quite a few selected LAs did not deliver
- coordinating LAs should be selected applying selection criteria rigorously, including: top scientific expertise in the relevant area, demonstrated management capabilities, time availability throughout the process, and ability to work productively with experts from disparate regions and disciplines
- flexibility is needed to bring in new LAs or CLAs during the process to fill gaps in knowledge or replace team members in case of poor performance.

The Co-Chairs of WG III provided further details on serious issues with the CLAs:[20]

From a management perspective, the non-performance of CLAs was an important issue. Having two CLAs per chapter had the advantage that practically always one was active, but the disadvantage was that the task sharing between the two was not always well arranged. In some cases, neither of the CLAs was taking sufficient responsibility to manage the process, and meet deadlines. In some cases, conflicts arose because of a lack of confidence in the (scientific or managerial) capabilities of CLAs by members of the writing team.

The TAR survived these weaknesses without major incident. However, serious shortcomings of the IPCC structure had been raised which the organization took no effective actions to correct, and, by the end of the decade, the IPCC was mired in a much bigger scandal.

Before all hell broke loose, the IPCC was oblivious to the impending danger and was clearly winning the fight for the hearts of the general public on the perils of climate change. By late October of 2009, just before Copenhagen, the AR4 was already a distant memory to the IPCC as they finalized the scoping document of the AR5 at the 31st Plenary Session[21] (October 2009). The *Earth Negotiations Bulletin* reported on the "buoyant mood."[22] Accolades were streaming in. For their contributions, the IPCC received the Nobel Peace Prize in 2007; quite an achievement for a scientific body. They shared the Prize with Al Gore (for his work on climate change and his documentary *An Inconvenient Truth*)[23]

> for their efforts to build up and disseminate greater knowledge about man-made climate change, and to lay the foundations for the measures that are needed to counteract such change ... Through the scientific reports it has issued over the past two decades, the IPCC has created an ever-broader informed consensus about the connection between human activities and global warming. Thousands of scientists and officials from over one hundred countries have collaborated to achieve greater certainty as to the scale of the warming. Whereas in the 1980s global warming seemed to be merely an interesting hypothesis, the 1990s produced firmer evidence in its support. In the last few years, the connections have become even clearer and the consequences still more apparent.

During the celebration of the twentiethth anniversary of the IPCC, on August 31, 2008, the Secretary-General of the United Nations, Ban Ki-moon had praised the IPPC:[24]

> The IPCC has a remarkable history of accomplishments including its First and Second Assessment Reports, which gave us the United Nations Framework Convention on Climate Change and the Kyoto Protocol, respectively.
>
> The critical role of the Fourth Assessment Report of last year, in putting to rest any left skepticism on whether climate change was real and happening; in elevating the discourse of response – no longer the debate – to the highest of political levels and public consciousness; and in fostering the breakthrough we managed to achieve in Bali is undeniable ... In the Bali Action Plan, Parties recognize that deep cuts in global emissions will be required to achieve the ultimate objective of the Convention and emphasize the urgency to address climate change as indicated in the Fourth Assessment Report of the IPCC.
>
> In this effort, we will continue to rely on our committed partner – the IPCC – as the key source of policy relevant science on climate change within the UN family.

And Yvo de Boer, the Executive Secretary of the UNFCCC, sent the following message:[25]

> The First Assessment Report in 1990 first alerted the world to the link between greenhouse gas emissions and global warming. With such alarming findings, based on the views of hundreds of experts, it was clear that the international community needed to respond. Negotiations were initiated and two years later, the United Nations Framework Convention on Climate Change was adopted and signed at the Earth Summit in Rio de Janeiro in 1992. In many ways, we owe the birth of the Convention to the work of the IPCC.
>
> The Second Assessment Report in 1995 had a similarly decisive impact. Its comprehensive assessment of climate change science strengthened the UNFCCC and opened the door for work towards the Kyoto Protocol, which was adopted in 1997.
>
> In a way, we are still riding the wave of public awareness that the Fourth Assessment Report created. The question is: will that wave carry us through to Copenhagen? Or will the negotiating process be working in a vacuum in 2009 because there will be no IPCC report?
>
> Reaching a strong agreed outcome in Copenhagen that matches up to the science is imperative. We need global public understanding, political commitment and strong science to result in a science-based political response.
>
> Public understanding is essential to secure political commitment. The public across the world and the climate change process itself need a scientific wake-up call for Copenhagen. My hope is that the IPCC can in some way help to supply that wake-up call.

A "wake-up call" was coming from the IPCC but not the type that de Boer had wished for. Ban Ki-moon had warned of increased pressures on the IPCC as the world prepared for COP 15 in Copenhagen in December 2009. The Secretary-General ended his speech by saying: "The world is waiting for results! The future is literally in our hands!"[26]

But as the IPCC was approaching the peak of its own Mount Everest, a scandal about the Himalayas would push them from their lofty heights.

In the AR4, there had been two errors in the Report of the WG II that caught media attention:

1. the area of land in the Netherlands that was below sea level
2. the rate of retreat of the Himalayan glaciers.

The former became known in the popular press as "Nethergate" and the latter as "glaciergate." While the problem of the Dutch sea levels was quickly dismissed, the glacier problem would not retreat so quickly.

"Nethergate"

In 2010, there were two inquiries directly into the IPCC. The first was carried out in the Netherlands.[27] A reporter for the weekly magazine *Vrij Nederland* (Free Netherlands) discovered an error ("Nethergate") in the AR4 of WG II; the phrase in question was:[28]

> The Netherlands is an example of a country highly susceptible to both sea-level rise and river flooding because 55% of its territory is below sea level where 60% of its population lives and 65% of its Gross National Product (GNP) is produced.

On January 28, 2010, the Dutch Parliament responded to the growing public pressure and requested Jacqueline Cramer (1951–), the Minister for the Environment, to investigate the two errors in the AR4, namely Nethergate and glaciergate.[29] The *Planbureau voor de Leefomgeving* (PBL; the Netherlands Environmental Assessment Agency[30]) was requested to examine Chapters 9–16 of WG II Report of the AR4 (i.e., the regional reviews which included the comments on the Netherlands' sea level and Himalayan glaciers), with a special focus on the glaciers.[31] The intent of the Dutch Parliament was to determine if "policymakers and the public at large could still trust the IPCC's key messages."[32] The PBL described their task as follows:[33]

> The investigation was conducted realising that climate science and policy-making are currently taking place in a new era, characterised by a high degree of politicisation, a much more dynamic interaction between science and public discourse, and vocal citizens who either want to know whether policy measures under discussion are all really necessary, or who are of the opinion that suggested measures do not go far enough.

They set out to answer the following questions:[34]

1. Are the summary conclusions on regional impacts well founded on the underlying chapters and literature references?
2. Are there errors in statements that have traveled from the scientific literature references and/or the main texts through to the summary conclusions?
3. If errors are found, do they affect the validity of these conclusions?
4. What recommendations can we derive from our investigation in order to further improve the quality of the assessment process for the Fifth Assessment Report (due in 2014)?

The PBL focused their efforts on the validity of summary conclusions. There were 32 statements in the *Executive Summaries*[35], four from each chapter (9–16) of the

AR4 of WG II, which appeared in the SPM of the Synthesis Report. The inquiry found no significant errors in the summary conclusions (one minor one was noted).[36] In the specific chapters under review, six errors were found, but none were included in the SPM or *Executive Summary*, and a few errors on specific references were also noted.[37] Among the errors in the body of the Report itself, only one "significant error" was found (besides the Himalayan glaciers and the Dutch sea levels):[38]

> [A] 50 to 60% decrease in productivity in anchovy fisheries on the African west coast was projected on the basis of an erroneous interpretation of the literature references. It appeared to be about a 50 to 60% decrease in extreme wind and seawater turbulence, with some effects on the anchovy population that were not quantified.

The fishy results, though, did not impact any of the summaries, and "anchovy-gate" escaped the wrath of the media.

A more disturbing revelation was that of the 32 summary conclusions in the *Executive Summary*, supporting evidence for seven of them could not be found within the body of the AR4 from WG II.[39] The BPL did not dismiss these statements but recommended that there be more transparency on how such conclusions are determined. The find, though, raised a more fundamental question – if the *Executive Summary* is not a summary of the body of the Report, what is it? The PBL recommended improvement on transparency, quality control, reducing errors and better review.[40] Overall, the PBL supported the AR4, and the study concluded:[41]

> Overall the summary conclusions are considered well founded and none were found to contain any significant errors.
>
> The Working Group II contribution to the Fourth Assessment Report shows ample observational evidence of regional climate-change impacts, which have been projected to pose substantial risks to most parts of the world, under increasing temperatures. However, in some instances the foundations for the summary statements should have been made more transparent. While acknowledging the essential role of expert judgment in scientific assessments, the PBL recommends to improve the transparency of these judgments in future IPCC reports.
>
> In addition, the investigated summary conclusions tend to single out the most important negative impacts of climate change. Although this approach was agreed to by the IPCC governments for the Fourth Assessment Report, the PBL recommends that the full spectrum of regional impacts is summarised for the Fifth Assessment Report, including the uncertainties. The PBL believes that the IPCC should invest more in quality control in order to prevent mistakes and shortcomings, to the extent possible.

When reading the Netherlands Environmental Assessment Agency (PBL) report, it is easy to forget the reasons behind the inquiry because of the detail of the analysis. The Dutch Parliament was not concerned about the minor nuisances of the

AR4 that were mentioned by the PBL; the simple question it wanted answered was whether "policymakers and the public at large could still trust the IPCC's key messages." And the thorough review confirmed that indeed they could.

What about the issue that was poignant to the Netherlands – the area below sea level? Nethergate[42] was not really a mistake by the IPCC, and the PBL[43] clarified the confusion on this matter – it was their fault.[44] And PBL accepted full responsibility for the mistake:[45]

> We acknowledge that this error was not the fault of the IPCC (Coordinating) Lead Authors or Co-Chairs. The error was made by a Contributing Author from the PBL, and the (Coordinating) Lead Authors are not to blame for relying on Dutch information provided by a Dutch agency. The lesson to be learned for an assessment agency such as ours, is that quality control is needed at the primary level of a literature assessment. This should be done by checking basic data, to the extent feasible, even when authoritative references are available. Of course, there are limitations to checking data, given the vast amounts of specific regional information.

There was more, however, to the retreating glaciers and further fallout for the IPCC was yet to come. Before the PBL report was issued, a broader inquiry was already underway by the InterAcademy Council.

"Glaciergate"

The long-standing Chair of WG I, John Houghton, had once attributed the success of the IPCC to not "overplaying what scientists knew."[46] But someone had "overplayed" in the AR4. This mistake led to a review of the IPCC requested by the UN itself.[47]

The response was grossly disproportionate to the gravity of the error itself. Within the tome of the Assessment Reports, we are talking about only one brief paragraph within the AR4 of WG II – the second paragraph of the short section 10.6.2 in Chapter 10:[48]

> Glaciers in the Himalaya are receding faster than in any other part of the world ... and, if the present rate continues, the likelihood of them disappearing by the year 2035 and perhaps sooner is very high if the Earth keeps warming at the current rate. Its total area will likely shrink from the present 500,000 to 100,000 km^2 by the year 2035 (WWF, 2005).

The important statement was highlighted in the chapter; the glacier issue is mentioned in a map of "hotspots" in Asia (Figure 2.1) and again appears in the Technical Summary of the AR4 of WG II (Box TS.6).[49]

Overall, the retreating glaciers of the Himalayas had a high profile within the AR4, but at least the issue is not mentioned in the critical SPM, although it had been there in earlier drafts.

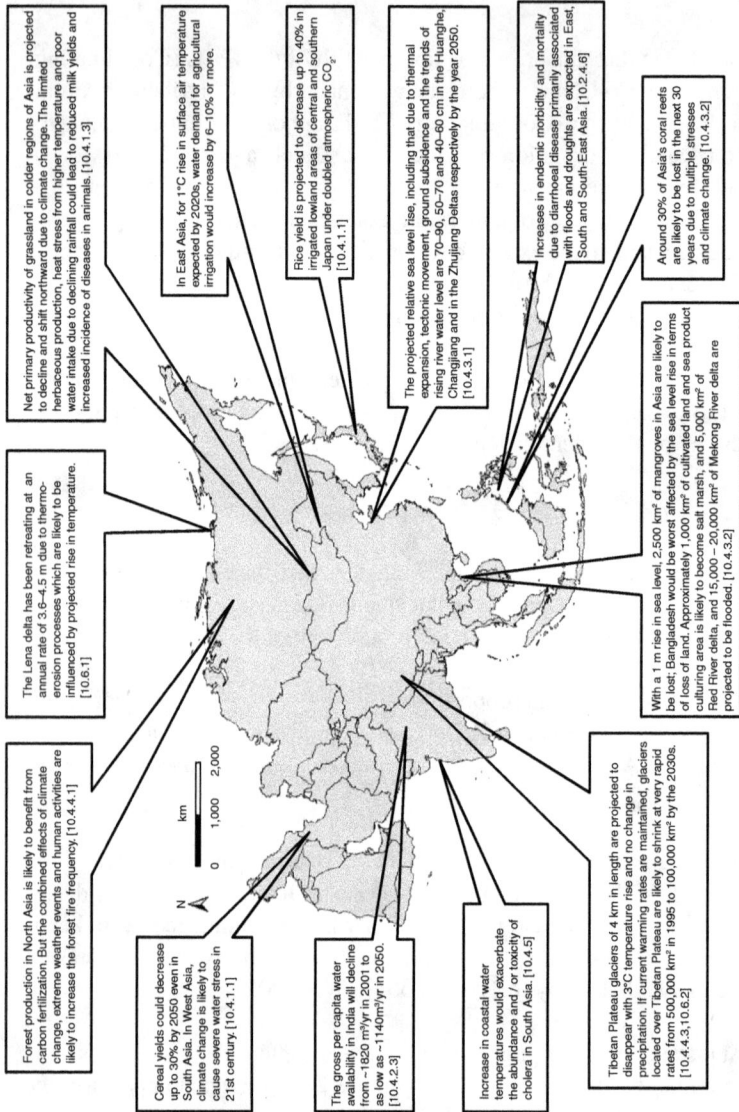

Forest production in North Asia is likely to benefit from carbon fertilization. But the combined effects of climate change, extreme weather events and human activities are likely to increase the forest fire frequency. [10.4.4.1]

The Lena delta has been retreating at an annual rate of 3.6–4.5 m due to thermo-erosion processes which are likely to be influenced by projected rise in temperature. [10.6.1]

Net primary productivity of grassland in colder regions of Asia is projected to decline and shift northward due to climate change. The limited herbaceous production, heat stress from higher temperature and poor water intake due to declining rainfall could lead to reduced milk yields and increased incidence of diseases in animals. [10.4.1.3]

In East Asia, for 1°C rise in surface air temperature expected by 2020s, water demand for agricultural irrigation would increase by 6–10% or more.

Rice yield is projected to decrease up to 40% in irrigated lowland areas of central and southern Japan under doubled atmospheric CO_2. [10.4.1.1]

The projected relative sea level rise, including that due to thermal expansion, tectonic movement, ground subsidence and the trends of rising river water level are 70–90, 50–70 and 40–60 cm in the Huanghe, Changjiang and in the Zhujiang Deltas respectively by the year 2050. [10.4.3.1]

Increases in endemic morbidity and mortality due to diarrhoeal disease primarily associated with floods and droughts are expected in East, South and South-East Asia. [10.2.4.6]

Around 30% of Asia's coral reefs are likely to be lost in the next 30 years due to multiple stresses and climate change. [10.4.3.2]

Cereal yields could decrease up to 30% by 2050 even in South Asia. In West Asia, climate change is likely to cause severe water stress in 21st century [10.4.1.1]

The gross per capita water availability in India will decline from ~1820 m³/yr in 2001 to as low as ~1140m³/yr in 2050. [10.4.2.3]

Increase in coastal water temperatures would exacerbate the abundance and / or toxicity of cholera in South Asia. [10.4.5]

With a 1 m rise in sea level, 2,500 km² of mangroves in Asia are likely to be lost; Bangladesh would be worst affected by the sea level rise in terms of loss of land. Approximately 1,000 km² of cultivated land and sea product culturing area is likely to become salt marsh, and 5,000 km² of Red River delta, and 15,000 – 20,000 km² of Mekong River delta are projected to be flooded. [10.4.3.2]

Tibetan Plateau glaciers of 4 km in length are projected to disappear with 3°C temperature rise and no change in precipitation. If current warming rates are maintained, glaciers located over Tibetan Plateau are likely to shrink at very rapid rates from 500,000 km² in 1995 to 100,000 km² by the 2030s. [10.4.4.3,10.6.2]

N

km

0   1,000   2,000

Figure 2.1. A Glaciergate error reported in the AR4 from the IPCC.

Source: IPCC, *Fourth Assessment Report, Working Group II, Impacts, Adaptations and Vulnerability,* 2007, p. 481; *Impacts, Adaptation and Vulnerability. Working Group II Contribution to the Fourth Assessment Report of the Intergovernmental Panel on Climate Change,* Figure 10.4., Cambridge University Press.
Note: The glacier section is in the seven o'clock position.

Indian government

During the government review process of the second draft, India had strongly objected to the comment appearing in the SPM.[50] The offending phrase was removed, and at the Plenary Session where the AR4 was approved, the Indian delegation went further and argued against the inclusion of paragraph 10.6.2 in the body of the Report,[51] but it passed anyway. India was not going to let the issue rest, which led to a clash between two of the most prominent actors in the Indian environmental scene – the Chairman of the IPCC, Rajendra Pachauri, and the Environment Minister of India, Jairam Ramesh. In an unprecedented move, the Indian government sponsored its own review on the Himalayan glaciers (*Himalayan Glaciers: A State-of-Art Review of Glacial Studies, Glacial Retreat and Climate Change*) by Vijay Kumar Raina.[52] The report was released in early November 2009, and it refuted the controversial glacier claims in the AR4.

There had been talk among the Indian scientific community, especially from the Wadia Institute of Himalayan Geology, that the IPCC Report was wrong about the Himalayan glaciers. Pallava Bagla[53] – Science Editor for NDTV (New Delhi Television) and Chief Correspondent, South Asia, for the journal, *Science* – had been looking into the story for two years, and the Raina report led him to go public. On November 9, 2009, Bagla broke the story on 24x7 News.[54] The network interviewed Pachauri for their broadcast, which was even more controversial than the AR4 itself. The Chairman was indignant about the charges. When the news broadcaster suggested that the IPCC should look at Raina's study, the response from the Chairman was dismissive; he stated that "they are totally wrong" and then uttered his famous comment, "this is voodoo science!" When the interviewer suggested that the IPCC discuss the issue with Jairam Ramesh, Pachauri stated: "I don't think he has any business questioning (the IPCC) … I will not sit down with the Minister (on this issue)" and then added, "he is a good friend of mine." In another interview appearing in the *Guardian*; a defiant Pachauri further criticized Raina and Ramesh:[55]

> We have a very clear idea of what is happening. I don't know why the minister is supporting this unsubstantiated research. It is an extremely arrogant statement … With the greatest of respect this guy [Raina] retired years ago and I find it totally baffling that he comes out and throws out everything that has been established years ago.

The derogatory criticisms and dismissive attitude by the Chairman of the IPCC are inappropriate under any circumstances, but were made even worse since Raina was right! However, Raina had climate skeptic leanings; he later was a contributor to an open letter to the UN denouncing the future threats of climate change.[56]

On November 13, 2009, Bagla reported on the issue in the journal, *Science*, accompanied by widespread support among glaciologists for the Raina report.[57]

Pachauri stated that he was "disappointed" that the prestigious journal had published the story.[58] And the BBC picked up the Bagla story on December 5.[59]

Besides the initial outbursts by Pachauri, the IPCC was strangely silent on the matter. No communications were made during the latter period of 2009. Only two months after the Bagla story broke did the IPCC finally respond with its famous "regret" statement in a short press release entitled *IPCC Statement on the Melting of the Himalayan Glaciers*:[60]

> It has, however, recently come to our attention that a paragraph in the 938-page Working Group II contribution to the underlying assessment refers to poorly substantiated estimates of rate of recession and date for the disappearance of Himalayan glaciers. In drafting the paragraph in question, the clear and well-established standards of evidence, required by the IPCC procedures, were not applied properly.
>
> The Chair, Vice-Chairs, and Co-chairs of the IPCC regret the poor application of well-established IPCC procedures in this instance. This episode demonstrates that the quality of the assessment depends on absolute adherence to the IPCC standards, including thorough review of "the quality and validity of each source before incorporating results from the source into an IPCC Report". We reaffirm our strong commitment to ensuring this level of performance.

In an interview by Bagla shortly afterwards, Pachauri was as defiant as before and stated:[61]

> So I cannot possibly be held accountable for all the lies that the media are writing about in a certain section of the U.K. press. I mean, if they are going to influence public opinion, I can assure you it is not going to last forever. I am absolutely convinced the truth will prevail in the end.

Pachauri was asked why the IPCC only expressed regret and did not apologize for the mistake. The Chairman used the question to criticize the Indian government:[62]

> We have made a mistake and we have admitted that. Our job is essentially to bring the science into our assessments from the best sources that exist. Look at the extent of the glaciology work that has been done in this country. It is pathetic. I mean, that is really where we need to come up with an apology.

A week later, on February 4, 2010, Jairam Ramesh, the Environment Minister of India, responded: "There is a fine line between climate science and climate evangelism. I am all for climate science but not for climate evangelism. I think people misused the IPCC report.[63]

The Mount Everest of controversies

By February 2010, glaciergate had morphed into a full-blown crisis for the IPCC. The IPCC had often received criticism from climate change skeptics, but glacier-gate had been taken to a new level as condemnations came, this time, from within its own ranks. In an unprecedented move, on February 11, the prestigious journal *Nature* hosted a forum that included members of the IPCC about the future of the organization, under the ominous but pithy title, "IPCC: Cherish It, Tweak It, or Scrap It":

- *Eduardo Zorita*[64] – CA of AR4 – mentioned the "failure of trust" and "flaws in its structure." He recommended the formation of an International Climate Agency, which would have 200 full-time scientific staff and would prepare biennial climate reports from peer-reviewed sources only.
- *Thomas F. Stocker*[65] – CLA of TAR and AR4 and Co-Chair of WG I for AR5 – defended the IPCC but stated that the rules and the procedures had to be followed. He did acknowledge the "recent controversies."
- *Jeff Price*[66] – LA of TAR and AR4 – questioned the process on how the CLAs and Review Editors were chosen. The best experts should be chosen and not be influenced by gender and geographical location. He also suggested more frequent and shorter reviews of peer-reviewed literature.
- *John R. Christy*[67] – LA of AR4 – recommended that the IPCC be removed from control of the United Nations. The problem is that governments often nominated only authors that were aligned to their biased views of climate change. To open up the process, Christy recommended a "Wikipedia-IPCC," where a select group of scientists would continuously update the information on climate change.
- *Mike Hulme*[68] – CLA, LA and Review Editor of TAR – recommended the dis-solution of the IPCC after AR5, and its replacement with three separate panels. The Global Science Panel would be similar to WG I, but the reports would be more frequent and much shorter; the second would be composed of various Regional Evaluation Panels; and the third would be the Policy Analysis Panel.

Earlier, on February 5, Hulme had also criticized the dismissive attitude of the Chairman: "Pachauri's haughty attitude helps explain why the controversy surrounding the mistaken claim – which, after all, is a rather minor piece of the picture of climate change impacts – is now filling newspapers, blogs and broad-cast media".[69]

And in an interview on February 10, Robert Watson also challenged Pachauri: "The mistakes all appear to have gone in the direction of making it seem like climate change is more serious by overstating the impact. That is worrying. The IPCC needs to look at this trend in the errors and ask why it happened."[70]

These were serious accusations by Pachauri's predecessor at the IPCC. Watson had gone as far as to have discussions with Al Gore to form a new scientific climate group, a "Wikipedia for climate change," to "supplement" the IPCC (similar proposals have been made by Neville Nicholls[71] and John Christy[72]).

US Environmental Protection Agency (EPA) inquiry

The EPA had never intended to investigate glaciergate, but it found itself in the wrong place at the wrong time. On December 7, 2009, just as the scandal (and that of climategate) was making headlines around the world, the Administrator of the US Environmental Protection Agency signed a statement that declared greenhouse gases to be a threat to public health. The EPA would have attracted a flood of comments by its declaration anyway, especially with there being such a strong skeptic lobby within the US, but the errors in the AR4, plus the emails from the CRU (climategate, *vide infra*), added a new dimension to this already intimidating process. The AR4 had been a prime reference for the EPA legislation, and the CRU allegations had led to accusations that the IPCC was not objective, and even corrupt.[73] The flood of comments was divided into 11 volumes; Volume 2, entitled *Validity of Observed and Measured Data*, handled accusations relating to the CRU and was supplemented by four Appendices:

1. Appendix A – Climate[74] Research Unit (CRU) Temperature Data Web Site
2. Appendix B – CRU Statement on Data Availability
3. Appendix C – United Kingdom Hadley Centre Statement on Release of CRU Data
4. Appendix D – Response of Keith Briffa to Stephen McIntyre.

In the companion reply document of three volumes, *Response to Petitioners*, the information in Volumes 1 and 3 was indirectly related to the IPCC:

Volume 1:
1. Validity of paleoclimate temperature reconstructions and related issues
2. Attribution of recent temperature trends and models
3. Validity of the HadCRUT temperature record
4. Validity of NOAA and NASA temperature records
5. Implications of new studies and data submitted by the petitioners.

Volume 3:
1. Approaches and processes used to develop the scientific support for the findings
2. Response to claims that the assessments by the USGRCP and NRC are not separate and independent assessments
3. Issues concerning the integrity of peer-reviewed literature
4. Issues concerning freedom of information act requests.

The EPA found no merit in the criticisms from the petitioners in Volumes 1 and 3.

Volume 2 was especially on the IPCC and the AR4, and comments were categorized by the EPA into two groups:

1. IPCC Errors – undermine IPCC findings and technical support for endangerment
2. IPCC policy agenda – not objective and impartial.

The following organizations were associated with the 34 accusations against the IPCC:

- Southeastern Legal Foundation – 56 percent (of the 34 accusations)
- Peabody Energy – 50 percent
- State of Texas – 32 percent
- Competitive Enterprise Institute – 24 percent
- Ohio Coal Association – 24 percent
- Coalition for Responsible Regulation – 15 percent.

Several petitioners had alleged that the science of the IPCC is "uncertain and/ or not credible."[75] The most comments (14 out of 15 accusations) in the first section on "IPCC Errors" came from the Southeastern Legal Foundation. What is the Southeastern Legal Foundation? The group describes itself as a "national constitutional public interest law firm and policy center that advocates limited government, individual economic freedom, and the free enterprise system in the courts of law and public opinion."[76] In other words, in the hidden language of the movement, they were a front for climate skeptics as documented by their criticisms and language; they searched the AR4 and the CRU emails looking for ways to attack the IPCC by interpreting any information in the worst light possible. Noticeable was the inflammatory language against the IPCC from this group: "their conclusions can be fake";[77] "fraudulently asserted";[78] "appears to be scientific fraud";[79] "IPCC horror stories";[80] and "so called 'science' of global warming."[81] The EPA in their response to the comments from the Southeastern Legal Foundation stated: "And they impugn the integrity of several scientists, on the basis of speculation not evidence, seemingly for no other reason than that they disagree with the conclusions of the assessment literature."[82]

The EPA summarized the section dealing with the "IPCC Errors" as:[83]

The small number of real or alleged errors in the IPCC report is not materially relevant for EPA's Endangerment Finding. Neither of the two errors that are verifiable (the percentage of the Netherlands below sea level and the projections for Himalayan glaciers) is relevant to the United States and neither is contained within EPA's endangerment record. Although it is unfortunate when any error occurs, two errors in a nearly 3,000-page study are not evidence of a complete breakdown of the IPCC's review process, contrary to petitioners' suggestions.

The remaining alleged errors or unsupported statements have not been shown to be erroneous and can be traced to findings in the gray literature from which they were derived. As discussed further in Subsection 2.2.4.4 in this volume and previously in Volume 1 of the RTC, the IPCC provides guidance on how and when to use gray literature, and petitioners do not demonstrate that the guidance was not followed. There is no de facto assumption that all gray literature is incorrect or suspect, and an examination of the instances raised by petitioners demonstrates that the specific allegations of petitioners are unfounded. Furthermore, the statements at issue drawn from the gray literature have no material relevance to conclusions supporting EPA's Findings, as explained in RTP Subsection 2.1.1.

   In conclusion, the evidence cited by petitioners does not undermine the overall credibility and reliability of the scientific conclusions relied upon and underlying EPA's Endangerment Finding.

Skeptics were also active with the second section on the "IPCC Policy Agenda" where criticism was more focused on the IPCC. The major contributor to this section (11/19 comments) was Peabody Energy, the world's largest private sector coal company. The accusations in this section were grouped by:

1. Inappropriate authorship and reviewer roles among IPCC personnel
2. Citing one's own papers compromises objectivity
3. IPCC is biased
4. IPCC lead authors forced consensus and suppressed dissenting views
5. Inappropriate collaborations among IPCC authors
6. Insufficient peer-review processes for the IPCC
7. IPCC authors manipulated deadlines for new authors
8. IPCC's use of gray literature.

After reviewing the comments, the EPA again supported the IPCC:

The evidence provided by the petitioners does not support their allegations that the IPCC's peer review and report development processes are designed inappropriately, or that they were inappropriately implemented in the development of the AR4.

The EPA legislation had become a magnet for the climate skeptic lobby and their frivolous attacks. US law invited such onslaughts and forced the EPA to defend itself and, in this case, the IPCC. The EPA verdict on the IPCC was that it was innocent of all charges, and the scientific integrity of the AR4 was sound.

   During 2010 a number of inquiries were underway into climate science, but there was one that was a bit different from the rest, one that had not come from skeptic politicians or even a nation; this review was requested by the United Nations. And another special aspect of this inquiry and its 100-page report was its cost: almost $1 million.[84] At $10,000 per page, the recommendations should have been important to the IPCC.

United Nations inquiry

The "regret statement" on January 20, 2012, had done nothing to relieve the anxiety of national governments over the growing controversy about glaciergate. On January 28 the Dutch government had launched their own inquiry (*vide supra*). The IPCC informed its member states on February 16 that it planned to set up an independent committee to examine its procedures,[85] but the response was too little too late. Official statements implied that the IPCC was initiating the process, but other reports indicated otherwise. On the 24th Pachauri was called to a special meeting of environment ministers in Bali organized by UNEP and the WMO. A report by *the Daily Telegraph* claimed that the concern was not over the mistakes in the AR4, but the manner in which Pachauri was handling the situation.[86] Three days later, Pachauri issued a statement about the Bali meeting:[87]

> But we recognize the criticism that has been leveled at us and the need to respond. While embarking on the preparation of its Fifth Assessment Report it was the intention of the IPCC that an independent committee of distinguished experts evaluate means by which IPCC procedures must be implemented fully and that they should also examine any changes in procedure that may be required.

Independent observers reported that Pachauri was not to be involved in the choice of the reviewers.[88] Yet the public announcement of the inquiry by the InterAcademy Council (IAC)[89] came from Ban Ki-moon and Pachauri.[90] Only a short while before Ban Ki-moon had stated:[91]

> we ought to celebrate the rigorous scientific tradition that underpins the work of the IPCC. Without the strong, peer reviewed science base and the policy relevance of that science which the intergovernmental process naturally engenders, the case for action on climate change would not be as unequivocal as it is today.

The review process was announced on March 10,[92] but not until April was a contract signed between UNEP and the IAC, and only in May had the 12 members of the expert review committee (chaired by Howard Shapiro, President Emeritus of Princeton University) been chosen. Besides their direct review of the IPCC documents and procedures, public input was also sought, which included more than 400 witnesses. The IAC politely described the reasons for the review:[93]

> In response to some sustained criticism and a heightened level of public scrutiny of the Fourth Assessment Report, the United Nations and IPCC asked the InterAcademy Council (IAC) to assemble a committee to review the processes and procedures of the IPCC and make recommendations for change that would enhance the authoritative nature of its reports.
>   Our review was undertaken amid a flurry of interesting, very public discussions surrounding aspects of IPCC's fourth assessment that raised concerns in

some quarters regarding the continuing credibility of the IPCC assessments themselves and the processes and procedures underlying them.

Their focus was not to assign blame but to strengthen the procedures and processes for the Assessment Reports. The review had three major chapters focusing on:[94]

2. Evaluation of IPCC's assessment process
3. IPCC's evaluation of evidence and treatment of uncertainty
4. Governance and management.

Among their recommendations were:[95]

- The IPCC should establish an Executive Committee[96] to act on its behalf between Plenary Sessions.
- The IPCC should elect an Executive Director to lead the Secretariat and handle day-to-day operations of the organization.
- The IPCC should encourage Review Editors to fully exercise their authority to ensure that reviewers' comments are adequately considered by the authors and that genuine controversies are adequately reflected in the report.
- The IPCC should adopt a more targeted and effective process for responding to reviewer comments.
- Each Working Group should use the qualitative level-of-understanding scale in its Summary for Policymakers and Technical Summary, as suggested in IPCC's uncertainty guidance for the Fourth Assessment Report.
- Quantitative probabilities (as in the likelihood scale) should be used to describe the probability of well-defined outcomes only when there is sufficient evidence.
- The IPCC should complete and implement a communications strategy that emphasizes transparency, rapid and thoughtful responses, and relevance to stakeholders.

The first Plenary meeting of the IPCC after the IAC review was the 32nd, in Busan, South Korea, on October 11, 2010. Pachauri started off the meeting:[97]

He noted that the past year had been a challenging period for the IPCC, but underscored that the independent review by the InterAcademy Council (IAC) had concluded that "the IPCC can claim many accomplishments to its credit," and that "the assessment process is successful overall." Noting the need to take action during this Session, Mr Pachauri emphasized a government-driven and transparent process to address the recommendations of the IAC Review of the processes and procedures of the IPCC.

Representatives of UNEP and WMO spoke next, and both were supportive of the IAC inquiry and the IPCC. Then followed the Executive Secretary of the UNFCCC, Christiana Figueres (1956–), who referred to the "confusion" of the past year, and that she supported the IAC Review. Figueres then added: "[T]he IPCC must increase the robustness and the quality of its assessments."[98]

What went wrong with 10.6.2?

The infamous section on glaciergate had appeared in the AR4 of WG II: Chapter 10: "Asia; 10.6. Case Studies; 10.6.2. The Himalayan Glaciers." The entire section in question, 10.6.2, contained only three short paragraphs. The first paragraph was just an introduction and contained no references. The third (and last) paragraph discussed data on specific glaciers and contained two references, one of which was to a 2002 paper of Syed Iqbal Hasnain, whose role in the controversy is discussed below.

The middle (second) paragraph was the problem:[99]

> Glaciers in the Himalaya are receding faster than in any other part of the world ... and, if the present rate continues, the likelihood of them disappearing by the year 2035 and perhaps sooner is very high if the Earth keeps warming at the current rate. Its total area will likely shrink from the present 500,000 to 100,000 km^2 by the year 2035 (WWF, 2005).

The IPCC Assessment Report system had failed; the LAs and CLAs had missed the obvious; the reviewers were ignored; and the Chairman of the IPCC acted recklessly in its defense. We will analyze the direct actions that contributed to this malfunction of the Assessment Report system.

The lost reference!

We will first drill down through the literature to find the ultimate source of the misleading information. The only reference in the second paragraph to 10.6.2 was a World Wildlife Fund (WWF) paper.[100] The relevant clause in the reference stated:[101]

> As discussed in the thematic introduction to this regional status review, there is particular concern at the alarming rate of retreat of Himalayan glaciers. In 1999, a report by the Working Group on Himalayan Glaciology (WGHG) of the International Commission for Snow and Ice (ICSI) stated: "glaciers in the Himalayas are receding faster than in any other part of the world and, if the present rate continues, the livelihood [sic] of them disappearing by the year 2035 is very high". Direct observation of a select few snout positions out of the thousands of Himalayan glaciers indicate that they have been in a general state of decline over, at least, the past 150 years.

The prediction that "glaciers in the region will vanish within 40 years as a result of global warming" and that the flow of Himalayan rivers will "eventually diminish, resulting in widespread water shortages" (*New Scientist* 1999; 1999, 2003) is equally disturbing.

When, in early 2010, the above statement was found to be in error, the WWF promptly issued a correction on March 1 at the beginning of their report:[102]

CORRECTION

On page 29 of the following report WWF included the following statement:

"In 1999, a report by the Working Group on Himalayan Glaciology (WGHG) of the International Commission for Snow and Ice (ICSI) stated: 'glaciers in the Himalayas are receding faster than in any other part of the world and, if the present rate continues, the livelihood [sic] of them disappearing by the year 2035 is very high.'"

This statement was used in good faith but it is now clear that this was erroneous and should be disregarded.

The essence of this quote is also used on page 3 in the Executive summary where it states: The *New Scientist* magazine carried the article "Flooded Out – Retreating glaciers spell disaster for valley communities" in their June 5, 1999 issue. It quoted Professor Syed Hasnain, then Chairman of the International Commission for Snow and Ice's (ICSI) Working Group on Himalayan Glaciology, who said most of the glaciers in the Himalayan region "will vanish within 40 years as a result of global warming".

This statement should also be disregarded as being unsound.

WWF regret any confusion this may have caused.

The WWF, itself, had not been the original source of the information as it had only cited references in the popular science magazine, *New Scientist*. Looking into the *New Scientist* article from 1999, it specifically reported:[103]

MELTING Himalayan glaciers are threatening to unleash a torrent of floods into mountain valleys, and ultimately dry up rivers across South Asia. A new study, due to be presented in July to the International Commission on Snow and Ice (ICSI), predicts that most of the glaciers in the region will vanish within 40 years as a result of global warming.

"All the glaciers in the middle Himalayas are retreating," says Syed Hasnain of Jawaharlal Nehru University in Delhi, the chief author of the ICSI report. A typical example is the Gangorti glacier at the head of the River Ganges, which is retreating at a rate of 30 metres per year. Hasnain's four-year study indicates that all the glaciers in the central and eastern Himalayas could disappear by 2035 at their present rate of decline.

The *New Scientist* article and WWF report quoted the ICSI (International Commission on Snow and Ice[104]) report, *Report on Himalayan Glaciology*, and its author, the Chairman of the ICSI Working Group on Himalayan Glaciology,

Syed Iqbal Hasnain; but the report, itself, did not include such information.[105] Bagla, who first broke the story on glaciergate, interviewed Dr Hasnain, who stated that the claims of Raina were "unscientific" and that the Indian government has an "ostrichlike attitude in the face of impending apocalypse."[106] When Fred Pearce, the author of the *New Scientist* article, heard about the controversy, he too contacted Hasnain; Pearce reported that Hasnain had told him that his original comments had been "speculation" on his part.[107] However, in another interview, Hasnain reported a different version of the story:[108]

> I have not given any date or year on the likely disappearance of Himalayan glaciers. I had simply told the *New Scientist* in an interview that the mass of the glaciers will decline in 40 years. The date (2035) was their invention. I was misquoted in the report.

I contacted Dr Hasnain, who was very open on the matter. In an email sent to me on July 1, 2011, he wrote:

> Fred Pierce [sic] of New scientist is well known alarmist journalist who has invented 2035 date and inserted in interview. Question is why IPCC authors quoted from popular science magazine and not use peer reviewed materials as it always claims?

In a situation that can only be described as unbelievable, there appeared to be no source for the information in the AR4! However, a forensic literature search by Fred Plitz uncovered the missing source of information used for 10.6.2; it had not been the WWF report or *New Scientist* but a *Down To Earth* article in 1999 (curiously, the WWF report had listed the *Down To Earth* article in the references, but it was not mentioned in the body of the report), which stated (bolded text is identical to 10.6.2):[109]

> **"Glaciers in the Himalaya are receding faster than in any other part of the world and, if the present rate continues, the likelihood of them disappearing by the year 2035** is very high," says the International Commission for Snow and Ice (ICSI) in its recent study on Asian glaciers. "But **if the Earth keeps getting warmer at the current rate**, it might happen much **sooner**," says Syed Iqbal Hasnain of the School of Environmental Sciences, Jawaharlal Nehru University, New Delhi. Hasnain is also the chairperson of the Working Group on Himalayan Glaciology (WGHG), constituted in 1995 by the ICSI. "The glacier will be decaying at rapid, catastrophic rates. **Its total area will shrink from the present 500,000 to 100,000 square km by the year 2035**," says former ICSI president V M Kotlyakov in the report Variations of snow and ice in the past and present on a global and regional scale.

The ICSI report and Dr Hasnain were again mentioned. In the lengthy series of misquoted information, the one common source had been Syed Iqbal Hasnain; yet no one from the IPCC had ever bothered to contact Hasnain, according to him. At the time, he would not have been hard to find for he was working at TERI for Pachauri. Another source quoted in the *Down To Earth* article was Vladimir Mikhaylovich Kotlyakov, who had been President of ICSI from 1987–91. However, the Kotlyakov paper did not apply specifically to the Himalayan glaciers and the year was 2350, not 2035; the 1996 article by Kotlyakov stated:[110]

> The extrapolar glaciation of the Earth will be decaying at rapid, catastrophic rates – *its total area will shrink from 500,000 to 100,000 [km²]; by the year 2350* [emphasis added]. Glaciers will survive only in the mountains of inner Alaska, on some Arctic archipelagos, within Patagonian ice sheets, in the Karakoram Mountains, in the Himalayas, in some regions of Tibet and on the highest mountain peaks in the temperature [sic, should read "temperate"] latitudes.

Apparently, someone had confused both the 2350 date, replacing it with 2035, and that the projection of Kotlyakov had applied directly to the Himalayans. But the real problem was not misquoted information in a popular science article. The serious errors were how such information had found its way into the AR4 and then, more importantly, how the wrong reference had been quoted.

However, the situation was much worse than this. In the first draft of the Assessment Report of the AR4, Section 10.6.2 had no references! Now, a charge of plagiarism could be made against the CA of 10.6.2, for the person had taken direct quotes without referencing them. Only during the Second Draft review[111] was this serious issue even detected. Clair Hanson of the TSU of the IPCC noticed that in the draft report of Section 10.6.2, there was only one reference.[112] Interestingly, this lone citation had been to a paper by Syed Iqbal Hasnain in 2002 in the third paragraph. The notorious second paragraph in question had not even been referenced originally. The omission did not appear to be a simple oversight, as the lack of citations was a problem over this entire section; in the first draft of 10.6.2–10.8.3, only three references had been mentioned.[113] A "review" with no references in an Assessment Report should have been a red flag to someone on the writing team. How could the first review have missed this? In the second review, a correction was made but this just made matters worse. On the advice of the writing team, a few more references were added. Only then does the erroneous WWF reference appear.

The writing team

Those accountable for this section are not directly identified by the IPCC. Unlike most reviews, the authors of different sections in IPPC Reports are not specifically identified. The contributors are grouped under Chapter 10; these members of the writing team, including those of the Bureau are:

Contributing Authors (Chapter 10) – Shiv D. Atri of India, Joseph Canadell of Australia, Seita Emori of Japan, Daidu Fan of China, Hui Ju of China,

Shuangcheng Li of China, Tushar K. Moulik of India, Faizal Parish of Malaysia, Yoshiki Saito of Japan, Ashok K. Sharma of India, Kiyoshi Takahashi of Japan, Tran Viet Lien of Vietnam, Qiaomin Zhang of China

Lead Authors (Chapter 10) – Yurij Anokhin of Russia, Batima Punsalmaa of Mongolia, Yasushi Honda of Japan, Mostafa Jafari of Iran, Congxian Li of China, Nguyen Huu Ninh of Vietnam

Review Editors (Chapter 10) – Daniel Murdiyarso of Indonesia, Shuzo Nishioka of Japan

Coordinating Lead Authors (Chapter 10) – Rex Victor Cruz of Philippines, Hideo Harasawa of Japan, Murari Lal of India, Shaohong Wu of China

Editors – Martin Parry, Osvaldo Canziani, Jean Pulutikov, Paul van der Linden, Clair Hanson (the latter three were from the Technical Support Unit of WG II)

Co-Chairmen of WG II – Martin Parry of England, Osvaldo Canziani of Argentina

Chairman of IPCC – Rajendra Pachauri of India

Contributing author

There were 13 CAs for Chapter 10, not a good omen. The author of 10.6.2 is not directly identified in the AR4. I requested information from the IPCC Secretariat in February 2012, but they had no formal records of who contributed to specific sections of previous Assessment Reports. The IAC was critical of the arbitrary process used to pick the CAs and that the best experts were not being chosen.[114]

The section in question is poorly written and improperly referenced by any standard. The statement in the AR4 is actually contradictory on its own, stating as it did that the glaciers in the Himalayas will shrink "from the present 500,000 to 100,000 km^2" and then that they would "disappear" by 2035; the glaciers cannot do both by 2035. The ambiguity of 10.6.2 is apparent to anyone who reads it. Making matters worse, both statements are wrong, as neither is occurring.

A more fundamental problem was that the WWF reference, itself, should have raised warning flags to the IPCC. The WWF is a credible organization, but the article is only a review. Original sources are the only true references, and the ones that needed to be validated. Even though the procedures of the IPCC allow for the use of gray literature in the Assessment Reports, the merit of the reference must be considered by the CA. How could a review reflect the merit of the original source? In this case, the paragraph was not even originally referenced. The situation was exacerbated by the WWF reference, added after the fact since it was not the source of the (lifted) quotes in 10.6.2.

Expert review

The IPCC system was set up to catch such mistakes, and the expert review was the only process within the IPCC to work as planned – to a point. The workload was enormous; for Chapter 10 of the AR4 of WG II alone, there were 526 expert comments and 145 government comments!

A glaring issue was the references, or lack thereof, as discussed above. An expert reviewer had also cautioned the IPCC that the information in the second paragraph of 10.6.2 was incorrect.[115] This reviewer, Georg Kaser, also recommended that the information needed to be cross-checked with the section of the Assessment Report of WG I (Chapter 4, Section 4.5) on the science of glaciology.[116] Had this been done, it would have raised more danger signs as this section of the report of WG I did not support the statements in the second paragraph of 10.6.2; in particular, Section 4.5 of WG I stated:[117]

> Whereas glaciers in the Asian high mountains have generally shrunk at varying rates (Su and Shi, 2002; Ren *et al.*, 2004; Solomina *et al.*, 2004; Dyurgerov and Meier, 2005), several high glaciers in the central Karakoram are reported to have advanced and/or thickened at their tongues (Hewitt, 2005), probably due to enhanced precipitation.

However, the response of the writing team to Kaser's suggestion was summarily dismissive: "This is based on reading of the literature, and so should be consistent with WG1 Chapter 4."

The assumption by the writing team was that Section 10.6.2 was compatible with the science reported by the glacier experts of WG I, which it was not. The outcome suggested that the Working Groups were working in silos,[118] and that cross-cutting issues were being poorly addressed.

There were more opportunities to catch the mistake. Hayley Fowler of Newcastle University brought attention to references that should be included in the third paragraph of Section 10.6.2[119] which were contradictory to the second paragraph:

> I am not sure that this is true for the very large Karakoram glaciers in the western Himalaya. Hewitt (2005) suggests from measurements that these are expanding – and this would certainly be explained by climatic change in precipitation and temperature trends seen in the Karakoram region (Fowler and Archer, J Climate in press; Archer and Fowler, 2004). You need to quote Barnett *et al.*'s 2005 *Nature* paper here – this seems very similar to what they said.

But the writing team recorded: "Was unable to get hold of the suggested references will consider in the final version."

The comment that references in standard journals could not have been retrieved is confusing. If the references had been included, the exaggerated comments made in the middle paragraph of 10.6.2 might have been avoided.[120] Again, there were no changes to the final draft.

In addition to poor science, the second paragraph is contradictory; the glaciers cannot both shrink to 100,000 km^2 *and* disappear by 2035, as mentioned above. The ambiguity in the wording was also identified in the review process; David Saltz of the Desert Research Institute at Ben Gurion University had made several useful suggestions – he questioned the retreat to 100,000 km^2 just after the paragraph had said that they were going to disappear.[121] The cryptic notes of the writing team stated: "missed to clarify this one." The IAC review of the IPCC noted the importance of the comment from Saltz, recognizing the contradiction and the lack of response by the writing team.[122]

Government reviewer comments were also made on 10.6.2. The relevance of the statement on the retreating Himalayan glaciers was recognized by the government reviewer from Japan, who suggested that the issue be included in the SPM, but this person also questioned how the likelihood was determined. This time the writing teams stated that the changes were made.[123] When the SPM was then reviewed by governments, Austria suggested less severe language[124] and India strongly objected.[125] In this case, the offending statement had been removed. Since concerns about this alarming statement had again been mentioned, this time in the SPM, the writing team should have been wary that something was amiss in Section 10.6.2. If the criticism was valid for the SPM, should it not have led to questions of leaving it in the body of the Report?

Lead authors

The review process had been set up by the IPCC to catch such mistakes. So far the process was working well, but then the wheels fell off as the Lead Authors did not change the text in response to the Expert Reviewers. The LAs are also responsible for the quality of the references used, especially the gray literature. As discussed above, the WWF reference is obviously not an appropriate reference for a number of reasons. Yet, it was still missed.[126] The lack of proper review of the quality of the reference is illustrated by the first draft, where no one even noticed that this critical passage had not been originally referenced! Did the LAs seriously read the first draft?

Pachauri laid the blame for the mistake on the LAs not following the procedures of the IPCC:[127]

> We have got to ensure that all our procedures are followed in letter and spirit and with a huge amount of due diligence. I will personally make sure that all the lead author teams that are going to work on the fifth assessment report and our special reports observe this scrupulously, go the extra mile in making sure that we don't use any information that is questionable. What has happened only highlights the importance of the procedures that we have established. If they had been followed, we wouldn't have got into this unfortunate error.

Review editors

Pachauri later stated that the Review Editor should have caught the mistake.[128] Especially considering the expert reviewers had identified many problems with 10.6.2, responsibility for the lack of proper follow-up must reside with the REs, Daniel Murdiyarso and Shuzo Nishioka. Along with the LAs, the REs make up the writing team handling the reviewer comments. While the LAs draft the Reports, the REs are to ensure that the review process was being followed and that comments were adequately addressed.[129] However, the LAs make the final decision on the wording of the draft. While the number of reviewer comments was massive, Section 10.6.2 was of particular importance and still should have been better handled by the REs.

Coordinating lead authors

Murari Lal was the CLA for the Asia section (Chapter 10) of the regional analysis of WG II[130] who had overall responsibility for the entry that became known as glaciergate. Dr Lal was a noted member of the Indian Institute of Technology in Delhi; UNEP reported on this noted contributor to the IPCC:[131]

> [Lal] is the only scientist from South Asia to have served the select group of authors for the four scientific assessment reports of climate change conducted thus far by the Intergovernmental Panel on Climate Change continuously for the past two decades.

When Lal had been informed of the Raina report, he too had rejected the accusation that the claims in 10.6.2 were wrong.[132] A newspaper had reported that an expert reviewer had written to Lal a few months before the AR4 was issued, warning that 10.6.2 was incorrect and should be removed:[133]

> Last week, Professor Georg Kaser, a glacier expert from Austria, who was lead author of a different chapter in the IPCC report, said when he became aware of the 2035 claim a few months before the report was published, he wrote to Dr Lal, urging him to withdraw it as patently untrue.

In the same newspaper story, Lal claimed he never received the message. I contacted Dr Kaser who stated that he never attempted to contact Lal directly and had used "official IPCC channels," but late into the process.

Pachauri publicly accused Lal of not following procedures. When confronted with the Chairman's comment, Lal responded: "We as authors followed them to the letter ... Had we received information that undermined the claim, we would have included it."[134]

Lal later stated that he was familiar with the WWF reference in 10.6.2:[135]

> We knew the WWF report with the 2035 date was "grey literature" [material not published in a peer-reviewed journal]. But it was never picked up by any of the authors in our working group, nor by any of the more than 500 external reviewers, by the governments to which it was sent, or by the final IPCC review editors.

His comment about the 500 external reviewers not identifying problems is incorrect, as we discussed above. The reference had not even been added until the external reviewers noticed that there had been no reference. Lal then blamed Hasnain: "We relied rather heavily on gray [not peer reviewed] literature, including the WWF report ... The error, if any, lies with Dr Hasnain's assertion and not with the IPCC authors."[136]

Lal missed the point; it was the responsibility of the CLA to determine whether a reference was appropriate or not. The assertion by Lal that the writing team had no responsibility for the mistake of 10.6.2 puts into question his understanding of the entire Assessment Review process.

Co-Chairs of WG II

Pachauri also laid the blame on the Co-Chairs of Working Group II, Martin Parry (who felt the whole affair was overblown)[137] and Osvaldo Canziani.[138] The WG Chairs were the official editors of the Assessment Report, and they chose the CLAs and LAs. The IAC stated:[139]

> The Working Group Co-chairs have significant influence and control over the assessment, leading the preparation, review, and finalization of their Working Group report. The importance of the Co-chairs makes it essential that they have the highest scientific and leadership credentials.

Chairman of the IPCC

Pachauri claimed that he had "absolutely no responsibility" for the mistake.[140] As already discussed, the Chairman openly blamed others: the Co-Chairs of WG II, Martin Parry and Osvaldo Canziani; the CLA of Chapter 10, Murari Lal; the unnamed Review Editor, and the unnamed Contributing Author.

Statements from Pachauri about 10.6.2 continued to be made which were inconsistent with the facts: "Well, we obviously made an error which we have corrected, and it's there on the website of the IPCC. We did that as soon as it came to our attention."[141]

"Soon!" Almost two years after Bagla had uncovered glaciergate, no correction had been made to the AR4! An *Errata* section has recently been available online for all the Working Group Reports of AR4. When examined in late September 2011, there had been a correction to 10.6.2. However, it was only an entry to the table immediately following the contentious second paragraph.[142] I brought this omission to the attention of the IPCC on October 17, 2011. The erroneous claim about the retreating Himalayan glaciers was finally added to the *Errata* section three days after my email! The IPCC did not respond to my email until January 23, 2012, when the Secretariat stated:

> Unfortunately, the IPCC did not have an error protocol that could address such an error until recently, however the "IPCC Protocol for Addressing Possible

Errors in IPCC Assessment Reports, Synthesis Reports, Special Reports or
Methodology Reports" was adopted by the Panel at its 33rd Session in Abu
Dhabi, 10–13 May 2011.[143]

Even then the document was not fully corrected. Figure 10.4 (see Figure 2.1
above) still highlighted the retreating glaciers quoting the original error in 10.6.2.
I informed the IPCC of the map figure error on October 14, 2012. A response
was received the following day that they had to consult with WG II; on April
17, 2013, the IPCC informed me that the correction had been made. A fix for
an already substantiated error, especially one with such a high profile, taking
another six months, demonstrates that the IPCC must tighten up their post-report
correction system. Such a correction should be made in days (at the most weeks),
not months.

One would think that Pachauri had been aware of the uproar over 10.6.2 before
the AR4 was released. And certainly, he would have been briefed on the details
after the scandal had broken. During the IAC review, Pachauri gave a presentation at their first meeting,[144] where he was asked what had gone wrong with glaciergate; the Chairman stated that the expert reviewer comments were "not very
specific in this regard" and that the information was "buried" in the Assessment
Report.

The facts again conflict with the testimony of the Chairman. The information
was not just "buried" in the report, but also appeared in the Technical Summary,
where it was highlighted as one of only ten points on Asia in "Box TS.6 The
Main Projected Impacts for Regions," and the reviewer comments were ample
to identify that something was wrong. More importantly, the contentious phrase
had even been added but then removed from the SPM. How is it possible for one
simple paragraph within the Assessment Report to stand out more? There were
also special reasons why Pachauri should have been well informed about Section
10.6.2; this was science in his backyard. When the Minister of the Environment
from his homeland had been so vocal in opposition, the Chairman would have
been expected to have noticed.

In defense of Pachauri, the climate change world was becoming more and more
like the lawless Wild West. And the Chairman of the IPCC was the only sheriff in
town. Pachauri would shoot first and worry about justice later in this intellectual
gunfight. We must also remember that Pachauri had a real paying job with TERI
and being Chairman of the IPCC was just a part-time, *pro bono* assignment.

The root cause of glaciergate

A major organizational weakness in the affair had been the review process.
However, there had been an excuse for this. On the WG II Report, there were 1,181
expert reviewers and 40,000 comments![145] Chapter 10 alone had 882 comments
from expert reviewers in the first draft (which identified no issues in 10.6.2) and
526 expert reviewer comments and 145 government reviewer comments on the
second draft.[146] As the Netherlands Environmental Assessment Agency stated:[147]

Realistically speaking, a thousand-page assessment by hundreds of authors involving thousands of reviewers conducted within a limited timeframe could hardly be expected to be free of errors. Therefore, it is to be expected that some inaccuracies, insufficiently justified statements or other irregularities, escape even the most thorough drafting and review procedures.

The IPCC review process was not set up to handle the onslaught – what organization would have been? In hindsight, the AR4 should have been delayed to provide more time for the review. With the time pressure, many of the comments were handled in a superficial manner as the writing team had to quickly skim through them all. Considering the obstacles that they faced, one is surprised that other mistakes had not slipped through.

The UNFCCC, itself, must shoulder part of the blame. The timing of the Assessment Reports was set more by the COP than the magnitude of the scientific task at hand. This was especially true for the AR4, where the IPCC was being threatened if the report was not ready in time for the Bali meeting of COP 13 (December 2007). Also, the demand for the sector analyses made matters worse. Each sector analysis was a mini-review of climate science, whose more general expertise resided in WG I. A more efficient handling of the important Sector Reports would have been more in the form of Special Reports, where information is simply taken from the full Assessment Report. This would have protected the science, especially from lesser experts trying to review gray literature. The IAC added: "The Panel should consider whether the regional assessment should be released significantly later than the sector assessment in order to devote as many high-quality resources as possible to these important issues."[148]

Even so, the review process failed for 10.6.2. During the expert review of the first draft, there were no comments; a shocking omission under the circumstances. The errors within the draft 10.6.2 are self evident if the section was read. One worries that the first draft review is too casual and only the second draft receives proper attention.

While the expert and government review processes generally worked well in the second draft, the actions of the follow-up writing team failed. The actual recording of the comments appeared to be well done. However, the outcome sheets filled out by the writing teams should be a check list on what needed to be done. No one had gone back to ensure that the items had been carried out according to the instructions. A related deficiency was no prioritization. Comments on the scientific integrity should have been highlighted, but became just a blur hidden by the multitude of minor corrections. The IAC highlighted this fact in their report:[149]

> With a tight schedule for completing revisions, authors do not always do an adequate job of revising the text, and Review Editors do not always require an explanation for rejected comments. In the case of the incorrect projection of the disappearance of the Himalayan glaciers, for example, some of the review comments were not adequately considered and the justifications were not completely explained (see Box 2.1). Although a few errors are likely to

be missed in any review process, stronger enforcement of existing IPCC procedures by the Review Editors could minimize their number. This includes paying special attention to review comments that point out contradictions, unreferenced literature, or potential errors; and ensuring that alternate views receive proper consideration.

A disturbing feature was that the comments of the writing team on the proposals of the expert reviewers were often dismissive and presumptuous. One worries that the tedious and unglamorous review process was becoming a chore rather than an integral part of the Assessment Report process. In other words, the writing teams simply wanted to get this behind them so that they could get the report out.

This situation is troublesome, as the work of the IPCC was supposed to be beyond reproach: "[T]he IPCC must at all times appear as the final, most credible and authoritative last word on all aspects of climate change."[150] So spoke Pachauri in regards to the AR4.

Who among the climate scholars had harmed the reputation of the IPCC? A shorter answer could be provided on who wasn't involved. No one/group associated with 10.6.2, except for the reviewers, had come close to doing their job properly, from the CA and Lead Authors to the Chairman of the IPCC himself. The scope of the problem from so many "experts" indicates that the individuals involved were not the root cause of the problems associated with 10.6.2. The failure throughout the organization just illustrated the weakness of the IPCC having no formal requirements for choosing individuals with the proper qualifications to handle the tasks at hand, which was clearly identified by the IAC:[151]

> Most important are the absence of criteria for selecting key participants in the assessment process and the lack of documentation for selecting what scientific and technical information is assessed. The Committee recommends that the IPCC establish criteria for selecting participants for the scoping meeting, where preliminary decisions about the scope and outline of the assessment reports are made; for selecting the IPCC Chair, the Working Group Co-chairs, and other members of the Bureau; and for selecting the authors of the assessment reports.

In short, this self-defined expert organization had no definition of an expert. While scholars and experts tend, at times, to be flippant about defining themselves, the IPCC is too complex an organization to allow such matters to simply happen. The results speak for themselves.

Among the four weaknesses of the IPCC discussed at the beginning of this chapter, the lack of qualifications and standard operating procedures was the only one under the direct control of the IPCC, and it was in this area that the IAC had made the majority of their recommendations. Here was an opportunity to fundamentally improve the operations of the IPCC. The process had started well. "Guidance notes" were developed for the Review Editors.[152] Members agreed that qualifications be established for the CLAs and LAs, as had been recommended by the

IAC.[153] However, the final recommendations of the Task Group were very weak on scientific qualifications; the only requirements for the CLAs and LAs were:[154]

1. scientific, technical and socio-economic expertise
2. geographical representation
3. a mixture of experts, with and without IPCC experience
4. gender balance.

Making matters worse, the wording was indefinite, as the members "shall aim to reflect."[155] In other words, there were no hard and fast rules as to the qualifications of the senior writing team members.

The IAC had warned:[156]

> However, no matter how well-constructed IPCC's assessment practices may be, the quality of the result depends on the quality of the leaders at all levels who guide the assessment process. It is only by engaging the energy and expertise of a large cadre of distinguished scholars as well as the thoughtful participation of government representatives that high standards are maintained and that truly authoritative assessments continue to be produced.

Of added concern was that there were no criteria for the leadership posts within the Bureau, which the IAC suggested should be corrected: "The IPCC should develop and adopt formal qualifications and formally articulate the roles and responsibilities for all Bureau members, including the IPCC Chair, to ensure that they have both the highest scholarly qualifications and proven leadership skills."[157] There had been "qualifications," if you can call them that, to be a Bureau member – "relevant scientific expertise." The IAC had recommended the strengthening of this requirement and that Bureau members had "both the highest scholarly qualifications and proven leadership skills." The proposal was discussed at the 32nd Session (October 2010), and the debate that ensued baffles the outside observer:[158]

> Saudi Arabia, with Argentina and China, noted that the current IPCC procedure for the selection of Bureau members is clear and opposed the second part of the recommendation on ensuring the highest scholarly qualifications and proven leadership skills, saying that it is too judgmental. However, Germany, Australia, Switzerland, the Netherlands, Denmark and Belgium agreed with the recommendation.

How could any member of the IPCC be against the "highest scholarly qualifications and proven leadership skills"? The issue could not be resolved at the session and was passed on to the Task Group on Governance and Management to report back at the 33rd Session.[159] Pachauri reminded the Task Groups of the "importance of geographical balance." Although the recommendations from the Task Group did not go as far as the IAC recommendation, they were a step in the right direction and were a vast improvement over the current mundane "relevant scientific expertise":[160]

Members of the Bureau should

a.   have recognised scientific, technical or socio-economic and other relevant academic qualifications to a higher degree level or the equivalent in
professional experience;

b.   have a relevant publications record;

c.   be recognised for their scientific integrity as experts by their peers.

The approval meeting on the action plans of the Task Groups was the 33rd Session
(May 2011) in Abu Dhabi;[161] inexplicably, the new qualifications eventually issued
were essentially the same as the old ones: "Members of the Bureau should have
appropriate scientific and technical qualifications and experience relevant to the
work of the Bureau, as defined by the Panel."[162] The only major change was from
"relevant" to "appropriate" scientific expertise. Despite the obvious importance,
the IAC recommendations on qualifications for the Chairman and other Bureau
members had been disregarded by the IPCC.

At the end of the 35th Session in June 2012, the IPCC was ready to move on
from glaciergate and the IAC review, as their "housekeeping matters" had been
sorted out:[163]

> Approximately a year and a half ago, the IPCC embarked on a journey to reform
> its processes and procedures, based largely on recommendations from the inde-
> pendent review by the Inter-Academy Council (IAC) launched in the aftermath
> of controversies surrounding the Fourth Assessment Report. The 35th session
> of the IPCC in Geneva took care of the few outstanding issues related to the
> review ... Many agreed that since the IPCC has now sorted out its housekeep-
> ing matters the Panel is now well prepared for the work and intense public scru-
> tiny ahead, as it enters the last stage of the Fifth Assessment cycle.

Even though the IAC had done an admirable job mapping the way forward, the
IPCC failed to grasp the opportunity. The root cause analysis discussed above had
found serious deficiencies across the organization. The all-encompassing nature
of the crisis suggested a systemic problem within the IPCC, especially with regard
to leadership. Leadership was lacking during the AR4 process of WG II, and completely collapsed afterwards. The major concern was not that within this massive
undertaking a mistake was made, but that denial permeated throughout the senior
ranks of the IPCC. No one was responsible! And the underwhelming response to
the IAC review itself just highlighted the failure of leadership within the IPCC.
At the end, there were no major changes in the leadership structure despite the
recommendations of the IAC report.

"Climategate"

In the history of climate change science, by an unbelievable coincidence, the two
greatest (alleged) scandals erupted within days of each other. The second turmoil also started in mid-November 2009, when a university website had been

hacked.[164] The victim was the Climatic Research Unit (CRU) of the University of East Anglia (UEA) in the UK. The CRU is a pioneering but small group formed in 1972 to study the past climate of the world, especially utilizing temperature readings for the recent past, and tree-ring (dendroclimatology) and other data as a proxy for earlier temperatures. They were best known for their historic graphs of average global and North American temperatures that appeared in the Assessment Reports.

Over 1,000 private emails from the past dozen years or so from the CRU had been released on the World Wide Web. Among the 1,073 emails involving over 150 authors, the largest number came from two prominent climate change scholars:[165]

- Philip D. Jones, the Director of the CRU, a CLA and a Reviewer of AR4 (Chapter 3) – 174 emails
- Michael Mann, the Director of the Earth System Science Center at Penn State University and a Reviewer of AR4 (a LA of TAR and collaborator with Jones and Briffa) – 140 emails.

Two years later, on November 22, 2011, a second wave of emails from the CRU was released on the internet.[166] The second release took place just before COP 17 (December 2011) in Durban, South Africa; the UEA stated: "[T]hese emails have the appearance of having been held back after the theft of data and emails in 2009 to be released at a time designed to cause maximum disruption to the imminent international climate talks."[167] The released snippet of messages was easily taken to mean that climate data were being exaggerated, or even fudged, and that there was an overt campaign to bar publication of the work of contrarian views. The topics of the controversial emails (representing about 0.3% of the total)[168] can be grouped into three principal allegations:

1. *Integrity* – manipulation and suppression of data, especially with regard to temperature reconstruction and related tree-ring analysis
2. *Censorship* – influencing editorial policy of scientific journals by badgering and threatening journal editors; a related issue was that they suppressed comments in the AR4
3. *Transparency* (including Freedom of Information Act/Environmental Information Regulations).

Jones, the director of the CRU, especially, was under personal attack, and the scientist was devastated by the accusations and public outcry. He allegedly even received death threats,[169] as did Mann. The FBI looked into death threats made against two unnamed participants in the scandal.[170]

The emails in question are not very professional and certainly did not portray Jones and Mann in a good light, but we must understand the circumstances that led to these messages being sent; most were part of a long-standing feud with some noted climate skeptics that had been going on for over a decade. Jones was clearly frustrated by the hounding he was receiving from this small, but vocal

Variations of the Earth's surface temperature for:

(a) the past 140 years

(b) the past 1,000 years

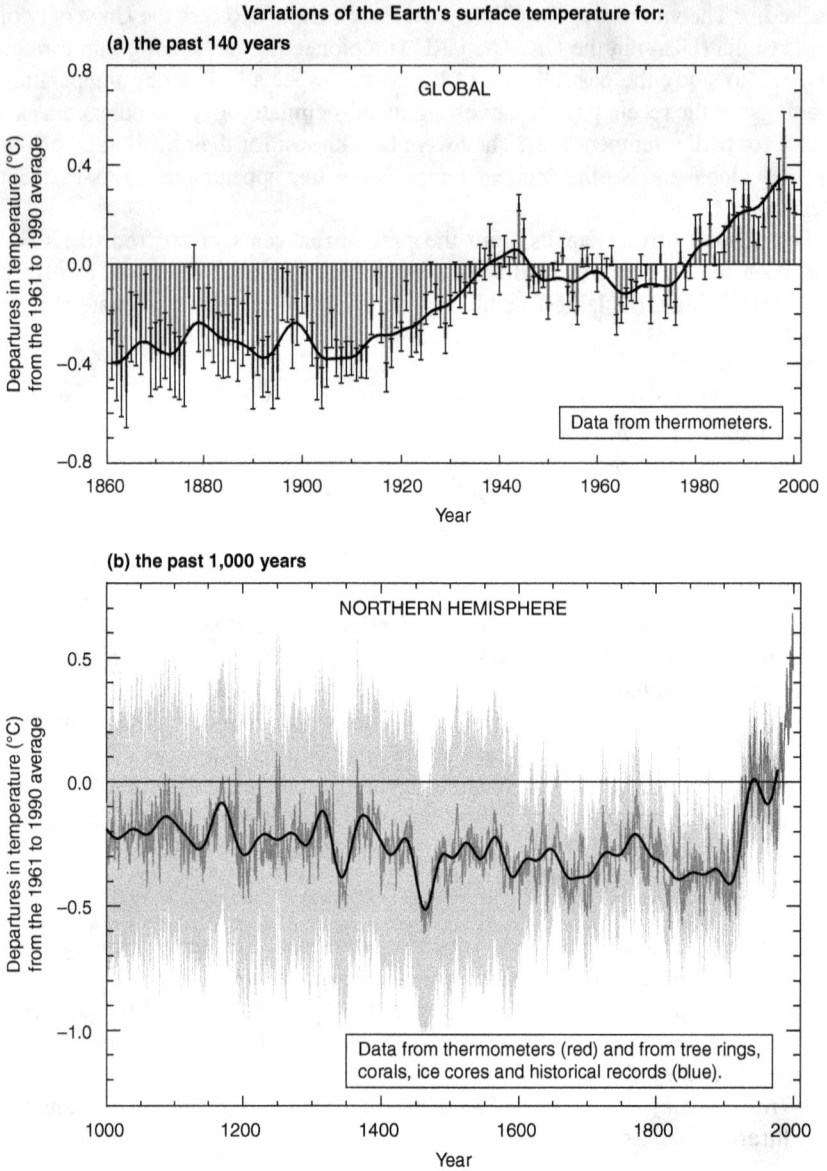

Figure 2.2. Historic temperatures.

Source: IPCC, *Third Assessment Report, Working Group I: The Scientific Basis*, p. 3; appears as Figure 1 in the SPM; in the *Technical Summary* as Figure 2, p. 26, and Figure 5, p. 29; in Chapter 2 as Figure 2.7c and Figure 2.20; *Climate Change 2001: The Scientific Basis. Contribution of Working Group I to the Third Assessment Report of the Intergovernmental Panel on Climate Change*, SPM, Figure 1, Cambridge University Press. The gray region shows the uncertainty as the 95 percent confidence range.

and persistent group. The language in his emails depict a dedicated researcher feeling persecuted by those who were deliberately attempting to undermine his research.[171] As a frustrated Jones wrote (in an email to the climate change skeptic Warwick Hughes): "We have 25 or so years invested in the work. Why should I make the data available to you, when your aim is to try and find something wrong with it?"[172] Most notable among this persistent group was Stephen McIntyre, a retired executive in the Canadian mining industry, and his sometime collaborator, Ross McKitrick, an economics professor at the University of Guelph.[173] Most of their discussions centered around one item, namely the iconic historic temperature graphs (Figure 2.2).[174] The "hockey stick" graph had first been published by Michael Mann[175], and McIntyre and McKitrick had identified errors back in the original 1998 paper of Mann.[176]

An entire book on the topic had been written by the noted skeptic Andrew Montford with the provocative title *The Hockey Stick Illusion: Climategate and the Corruption of Science*. And Mann came out with his own memoirs in 2012, *The Hockey Stick and the Climate Wars: Dispatches from the Front Line*. The science behind the graph has generally been supported by the scientific community at large, but the skeptic movement has focused on this allegedly weak link within the Assessment Reports.

Almost overnight, the information from the blogosphere had jumped to mainstream media. The BBC,[177] the *Guardian*[178] and the *Daily Telegraph*, where the term "climategate" was used[179], reported the story on November 20, 2009, and it had been picked up by the rest of the world's media the following day.[180] And on December 4, the story was released in the prestigious magazine, *Science*.[181] Robert Watson, the former Chair of the IPCC and Visiting Professor to the UEA, declared: "[T]he media has certainly portrayed the UEA issue as a crisis, so I think to the public it has been portrayed as a crisis."[182] The climate skeptics, of course, had a field day. They were now headliners among the popular press. On November 29, 2009, the geologist and noted climate skeptic Don Easterbrook stated:[183]

> Legitimate scientists do not doctor data, delete data they don't like, hide data they don't want seen, hijack the peer review process, personally attack other scientists whose views differ from theirs, send fraudulent data to the IPCC that is used to perpetuate the greatest hoax in the history of science, provide false data to further legislation on climate change that will result in huge profits for corrupt lobbyists and politicians, and tell outright lies about scientific data.

The views of climate skeptics were one thing, but prominent members of the scientific community, including some members of the IPCC, added fuel to the fire as the defenseless members of the CRU were being prepared to be burned at the stake. The *New York Times* reported the comments of the outspoken climate scientist Mike Hulme (also from the University of East Anglia, and who had worked for the CRU until 2000), who brought the IPCC into the debate:[184]

But this episode might signify something more in the unfolding story of climate change. This event might signal a crack that allows for processes of re-structuring scientific knowledge about climate change. It is possible that some areas of climate science has become sclerotic. It is possible that climate science has become too partisan, too centralized. The tribalism that some of the leaked emails display is something more usually associated with social organization within primitive cultures; it is not attractive when we find it at work inside science.

It is also possible that the institutional innovation that has been the I.P.C.C. has run its course. Yes, there will be an AR5 but for what purpose? The I.P.C.C. itself, through its structural tendency to politicize climate change science, has perhaps helped to foster a more authoritarian and exclusive form of knowledge production – just at a time when a globalizing and wired cosmopolitan culture is demanding of science something much more open and inclusive.

The first official reaction on the CRU emails from the IPCC came on December 4, 2009, when Pachauri stated that the IPCC was planning to investigate the allegations.[185] Even though Jones and Mann had long been valued contributors to the Assessment Report process, the IPCC quickly distanced itself from the growing controversy; in mid-December, Pachauri added:[186]

In every field of endeavour, there might be a few "black sheep" who deviate from ethical practices. But that does not mean the whole system can be labelled as being dishonest or fraudulent on the basis of one particular incident, even if it was true … I'm not aware of the actions of individuals. But IPCC is an extremely transparent body. Everything we do is in the public domain.

He later added that the emails "represent nothing more than private communications, private airing of anguish or anger or emotion. It was indiscreet."[187]

UK government inquiries

The email scandal at the UEA led to a British government inquiry into the matter, but this had not been the first British political intervention into climate science. A few years after the TAR had been issued, the skeptic movement in the UK became more prominent. Their activity may have been spurred on by the comments of the Chief Scientific Adviser to the UK government, Sir David King, who stated in early 2004: "Climate change is a far greater threat to the world than international terrorism."[188] Later in the year, the Scientific Alliance joined up with the George C. Marshall Institute, now being led by William O'Keefe, who had been Chairman of the GCC, to prepare a document casting doubt on climate change science.[189]

A hostile inquest against the IPCC by the Economic Affairs Committee of the UK House of Lords followed in July 2005. Most of the comments of the Committee sounded like rhetoric from climate skeptics:[190]

> We have some concerns about the objectivity of the IPCC process, with some of its emissions scenarios and summary documentation apparently influenced by political considerations.
>
> There are significant doubts about some aspects of the IPCC's emissions scenario exercise, in particular, the high emissions scenarios. The Government should press the IPCC to change their approach.
>
> There are some positive aspects to global warming and these appear to have been played down in the IPCC reports; the Government should press the IPCC to reflect in a more balanced way the costs and benefits of climate change.

Their report began by providing a reasonable overview of the climate change situation, but soon degenerated into a soapbox for the views of climate change skeptics. Chapter 2 of the report, for example, was entitled "The Uncertain Science of Climate Change." They presented the views of Richard Lindzen of MIT, a noted climate change skeptic. Following Lindzen's comments, they criticized the IPCC for not reporting on the contrarian point of view.[191]

> We recognise that there is a strong majority view on climate change. Majorities do not necessarily embody the truth, but we note that major associations of scientists have adopted similar positions. The IPCC tends to be the focus of the majority view which has been confirmed by the Royal Society, and by the US National Academy of Sciences, the American Meteorological Society, the American Geophysical Union and the American Association for the Advancement of Science. Despite this, it is a concern that the IPCC has not always sought to ensure that dissenting voices are given a full hearing. We document these concerns later in the Report.

The skeptic bent of the inquiry revealed itself again when the report went on to discuss the "hockey stick" debate.[192] They specifically mentioned the evidence of McIntyre and McKitrick, both leading climate skeptics, who, for the past decade, had been trying to destroy the credibility of this high-profile graph (*vide supra*). Then another skeptic, Paul Reiter, was called upon to criticize the IPCC.[193]

Climate change economics was highlighted in Chapter 5 on the costs of mitigation, and Chapter 6 on the cost of impacts. In the latter analysis, the Lords suggested in bold that the IPCC was attempting to conceal that the costs of impacts were relatively small compared to the costs of mitigation.[194] In the conclusion of the study, a section dealt with the IPCC process:[195]

> 171. We can see no justification for an IPCC procedure which strikes us as opening the way for climate science and economics to be determined, at least in part, by political requirements rather than by the evidence. Sound science cannot emerge from an unsound process (para 111).

172. The IPCC Summary for policy makers says that economic studies underestimate damage, whereas the chapter says the direction of the bias is not known (para 114).

173. We are concerned that there may be political interference in the nomination of scientists to the IPCC. Nominees' credentials should rest solely with their scientific qualifications for the tasks involved (para 116).

174. The IPCC process could be improved by rethinking the role that government-nominated representatives play in the procedures, and by ensuring that the appointment of authors is above reproach. At the moment, it seems to us that the emissions scenarios are influenced by political considerations and, more broadly, that the economics input into the IPCC is in some danger of being sidelined. We call on the Government to make every effort to ensure that these risks are minimised (para 118).

The British government generally dismissed the recommendations and allegations in the report.[196]

British House of Commons Science and Technology Committee first inquiry

An investigation of the CRU emails was initiated by the House of Commons (UK) Science and Technology Committee.[197] On December 1, 2009, the Chair of the Committee, Phil Willis (1941–), wrote to the Vice-Chancellor of the University of East Anglia (UEA), Edward Acton (1949–), about the CRU emails; specifically:[198]

- what had taken place
- the steps that had been taken to investigate the allegations and to test the integrity of the data held and used by CRU
- how CRU justified its commitment to academic transparency
- how the Vice-Chancellor proposed to restore confidence in CRU and its handling of data.

However, the Committee was not particularly impressed with the response of the University, considering "the weight UEA attached to restoring its reputation … in contrast to the other issues raised such as investigating the truth about the allegations made against the CRU."[199] The UEA then announced the formation of the Muir Russell review on December 3, and later another one – Lord Oxburgh's Science Assessment Panel – on March 22, 2010. The Muir Russell report focused on potential malpractice on the part of the scientists, and the Oxburgh report examined the merits of the science itself. The scope of the inquiries included the University of East Anglia, the Climatic Research Unit (CRU), Phil Jones, and the IPCC.[200]

The Science and Technology Committee held its own investigation. On January 22, 2010, the Committee first called for evidence and on March 31, their report was issued. The House of Commons inquiry laid out the main issues and the position of both sides, but not much more; the detailed investigation was left to the Oxburgh and Muir Russell inquiries. The Committee had issued its report quickly because of the upcoming general election.[201] They would return to the issue afterwards.

University of East Anglia – Independent Climate Change E-mails Review (Muir Russell) inquiry

On December 3, 2009, the *Independent Climate Change E-mails Review* (ICCER), under the direction of Sir Muir Russell (1949–), was announced, and the report was completed on July 7, 2010. Although called for by the University of East Anglia, the House of Commons Science and Technology Committee scrutinized the composition and operation of the Muir Russell committee.[202]

Jones had rejected a peer-reviewed reference during the Assessment Report process. However, according to the Muir Russell review, the IPCC does not simply accept all peer-reviewed references; references must meet the standards of the writing team, and the one in question did not.[203] Russell's conclusion, though, was inconsistent with an earlier statement made by the IPCC on peer-reviewed articles – made on December 4, 2009, the day after the Muir Russell study had been announced (emphasis added):[204]

> IPCC relies mainly on peer reviewed literature in carrying out its assessment and follows a process that *renders it unlikely that any peer reviewed piece of literature, however contrary to the views of any individual author, would be left out.* The entire report writing process of the IPCC is subjected to extensive and repeated review by experts as well as governments. Consequently, there is at every stage full opportunity for experts in the field to draw attention to any piece of literature and its basic findings that would ensure inclusion of a wide range of views. There is, therefore, *no possibility of exclusion of any contrarian views, if they have been published in established journals or other publications which are peer reviewed.*

The above was part of a general statement by the IPCC on the CRU email scandal.

Despite the conflicting statements, the exhaustive investigation found no major improprieties on behalf of Jones and the CRU. There was no evidence that the scientific credibility of the AR4 had been compromised by any of the implications arising from the CRU emails.[205]

University of East Anglia – Scientific Assessment Panel (Oxburgh) inquiry

The Scientific Assessment Panel (SAP), under Ernest Ronald Oxburgh (1934–), was announced by the UEA to carry out "an independent external reappraisal of the science in key CRU publications – to complement the ICCER."[206] The House of Commons Science and Technology Committee was not sure that the investigation should have been broken into two independent investigations, and it had suggested that Muir Russell and Lord Oxburgh coordinate their reviews to ensure that all areas were covered and that there was no overlap.[207]

Oxburgh started his investigation on March 22, 2010, and the report was issued in only a few weeks on April 14. Much of the review was on the tree-ring data used in the hockey stick graph (*vide supra*). They concluded the CRU could have been better at statistics,[208] but that they had acted with scientific integrity.[209] The Oxburgh report did have a small criticism of the IPCC, in that they did not provide a proper description of the uncertainties in the CRU data.[210]

British House of Commons Science and Technology Committee second inquiry

In the fall of 2010, with the general election behind them and the two UEA inquiries completed, the Science and Technology Committee met again, not to review the CRU but the inquiries themselves; Lord Oxburgh and Sir Muir Russell would be called upon to defend their reviews. The questioning was generally aggressive in nature, especially that of Graham Stringer (1950–), the Labour Party member from Blackley and Broughton, and Stephen Mosley (1972–), the Conservative Party member from Chester.

On September 8, 2010, Lord Oxburgh sat before the Committee.[211] Questions were raised on how the Oxburgh inquiry (SAP) took three weeks but the Muir Russell report (ICCER) took seven months. Oxburgh explained that he only had a limited remit and a deadline to meet. The Committee was not convinced and remained concerned about the brevity of the Oxburgh investigation:[212]

> The disparity in length between the SAP and ICCER reports is striking. When compared to the ICCER, the SAP report – a mere five pages – reads like an executive summary, with none of the detail of the ICCER. From Lord Oxburgh's evidence to us, the report does not appear to explain the detailed work carried out by the SAP. That in itself does not invalidate the SAP report but it does foster an impression that it was not as thorough as the ICCER and was produced quickly in an attempt to be helpful to UEA.

Concerns were also made on the transparency of the inquiry.

Next, on October 27, 2010, the investigation continued with Muir Russell and two senior administrators from the University of East Anglia, Edward Acton and Trevor Davis.[213] The Committee was a bit miffed that Muir Russell had not exactly followed their instructions, but it accepted his explanations. They also expressed reservations about transparency during the Muir Russell inquiry.[214]

The second review of the Committee found fault with both previous inquiries:[215]

> Lord Oxburgh said that CRU was not able to make accurate reconstructions in every case, particularly of old material. Professor Davies from UEA confirmed this but said CRU scientists would be able to do this given a number of weeks. This is precisely the sort of work we would have expected the Scientific Assessment Panel to conduct – had it been less concerned about

rushing to publish its report – during its inquiry into methodologies and the integrity of research at CRU.

We are concerned that the *Independent Climate Change E-mails Review* did not fully investigate the serious allegation relating to the deletion of e-mails. We find it unsatisfactory that we are left with a verbal reassurance from the Vice-Chancellor that the e-mails still exist. On the basis of the ICO's announcement made on 7 July 2010, it is reasonable to conclude that there was a breach of EIR by a failure to provide a response within 20 working days. On the allegation that e-mails were deleted to frustrate requests for information, a firm conclusion has proved elusive. UEA have accepted that there were weaknesses in their system, and in pockets of their culture, for dealing with requests for information. We are pleased that they are working towards rectifying this.

There was another inquiry of the inquiries, but not from government this time. Climate skeptics felt excluded from the reviews, and the Muir Russell report had been especially criticized by this group.[216] One of the more high-profile skeptic groups in the UK was the Global Warming Policy Foundation (GWPF), which had commissioned Andrew Montford, the author of *The Hockey Stick Illusion: Climategate and the Corruption of Science*, to review the UK probes. His main criticism was that the skeptics directly mentioned in the CRU emails were ignored by the inquiries, and many of the reviewers were prominent supporters of climate change so that the inquiries were biased in favor of the CRU.[217] Similar criticisms were raised by Montford regarding the Pennsylvania State University inquiry[218] (*vide infra*).

Assessment of the British inquiries

Although the investigations in the UK were somewhat independent of each other, the British government, through the Science and Technology Committee was carefully controlling all of them. And the second review of the Committee was actually an inquiry of the inquiries, namely that of Lord Oxburgh and Sir Muir Russell. The important conclusions were that Jones and the CRU were innocent of any wrongdoing.

The First Inquiry of the Committee had also largely dismissed the allegations with one exception – transparency by not sharing information.[219] Jones had refused to release raw data, but this was common practice, which the Committee recommended should change:[220]

Reputation does not, however, rest solely on the quality of work as it should. It also depends on perception. It is self-evident that the disclosure of CRU e-mails has damaged the reputation of UK climate science and, as views on global warming have become polarised, any deviation from the highest scientific standards will be pounced on. As we explained in chapter 2, the practices and methods of climate science are a key issue. If the practices of CRU are found to be in line with the rest of climate science, the question would arise whether climate science methods of operation need to change. In this event we would recommend that the scientific community should consider changing those practices to ensure greater transparency.

The second inquiry of the Science and Technology Committee expressed similar concerns:[221]

> While we do have some reservations about the way in which UEA operated, the SAP review and the ICCER set out clear and sensible recommendations. In our view it is time to make the changes and improvements recommended and with greater openness and transparency move on.

The criticism on openness of the CRU is a serious one, as science, especially publicly funded science, is not supposed to operate within a cloak (or lab coat) of secrecy. The problem was not just the CRU, but climate science in general.[222]

Jones and Mann had been abused, harassed and insulted by skeptics for more than a decade. Why would anyone volunteer to share information under these circumstances? The Muir Russell report understood the reluctance to release information in these cases but still supported the sharing of information:[223]

> More generally the majority climate scientists appear to have been united in their defence against criticism. Whilst perhaps understandable, given the nature and methods of criticism, some of which impugns their personal integrity as well as challenging their work, this may have blinded some CRU members to the possibility of merit therein. Such denial then fuelled yet further antagonism.

The ethical question of transparency within scientific circles is an important one. In principle, scientists in public institutions should and must go out of their way to share information with other scientists in their field. Scientists must promote the advancement of knowledge, and transparency is a mechanism in this regard. When a request is from another scientist in the field that is one thing, but from others the obligation to supply information is not so obvious. In this case though the sharing of data went beyond a purely ethical question, as the Freedom of Information Act (FOIA) had been violated. On the issue raised by the Committee of violating the FOIA, Jones was not directly blamed, as responsibility was placed with the University of East Anglia.[224] Of note, in a request from a climate change skeptic, the University refused on March 28, 2011 to release information.[225] The person in question was none other than the infamous Steven McIntyre.

Apart from breaking the letter of the law, a far greater concern should have been the threat of the FOIA to science, especially climate science.[226] Responding to such inquiries can be time consuming and takes people, including scientists, away from more pressing duties. In a concerted campaign, requests can overwhelm researchers. There must be a limit to how much time and effort scientists can be expected to divert from their research to handle frivolous or even malicious requests. Evidence of such an organized campaign was uncovered by the Muir Russell investigation. The CRU would normally receive a few requests under the

FOIA each year. However, in late July 2009, 60 were received in one week![227] The legal responsibility of sharing information under the FOIA in such situations must be re-examined in this new and unfortunate context. The House of Commons Committee ignored how the FOIA was being abused.

Furthermore, their investigation never mentioned the rights of Jones and his colleagues to freedom of speech, academic freedom and privacy. At least the Committee's investigation could have been more balanced by considering the violation of rights of leading British scientists, the waste of taxpayers' money being spent on suspicious requests under the guise of the FOIA and the criminal act of hacking the email system of the UEA.

Pennsylvania State University inquiry

As the University of East Anglia was organizing its investigations, the Pennsylvania State University, the home of Michael Mann, was doing the same. By November 22, 2009, Penn State had already been bombarded with accusations about the email scandal of the CRU from State and Federal politicians.[228] On November 29, the University informed Mann that he was being investigated for research misconduct (policy RA-10),[229] and the investigation began the next day. A review of more than 1,000 CRU emails was undertaken, and Mann was connected to over 300 of them. The University grouped the charges into four allegations:

1. Did you engage in, or participate in, directly or indirectly, any actions with the intent to suppress or falsify data?
2. Did you engage in, or participate in, directly or indirectly, any actions with the intent to delete, conceal or otherwise destroy emails, information and/or data, related to AR4, as suggested by Phil Jones?
3. Did you engage in, or participate in, directly or indirectly, any misuse of privileged or confidential information available to you in your capacity as an academic scholar?
4. Did you engage in, or participate in, directly or indirectly, any actions that seriously deviated from accepted practices within the academic community for proposing, conducting, or reporting research, or other scholarly activities?

The committee interviewed Mann on January 12, 2010, and, on January 26, the preliminary investigation found "no substance" in the first three, but the fourth allegation required further investigation.[230] This latter review, which began in April 2010, also found "no substance" to this allegation.[231]

National Science Foundation inquiry

Penn State had informed the National Science Foundation (NSF) of its investigation and findings.[232] The Office of the Inspector General of the NSF did not find that the University had provided enough evidence and,[233]

[w]e wrote to the University, requesting an extensive amount of documentation related to its investigation, including copies of all documentation the
committees used in their assessments, copies of all interview transcripts, and
specific transcripts or memorandums about certain conversations to which
the report referred. We also asked the University to address several questions
including:

1. How the inquiry committee determined that there was no substance to
    the first three allegations,
2. How both committees selected the individuals they interviewed,
3. How the investigation committee verified certain statements by the
    Subject and other individuals interviewed,
4. Evidence used by both committees to determine accepted practices, and
5. How the investigation committee reconciled differing viewpoints
    expressed by interviewees regarding data sharing expectations.

They then carried out their own investigation on the accusation of research misconduct and concluded:[234]

[N]o direct evidence has been presented that indicates the Subject fabricated
the raw data he used for his research or falsified his results. Much of the current debate focuses on the viability of the statistical procedures he employed,
the statistics used to confirm the accuracy of the results, and the degree to
which one specific set of data impacts the statistical results. These concerns
are all appropriate for scientific debate and to assist the research community
in directing future research efforts to improve understanding in this field of
research. Such scientific debate is ongoing but does not, in itself, constitute
evidence of research misconduct.

Lacking any direct evidence of research misconduct, as defined under the
NSF Research Misconduct Regulation, we are closing this investigation with
no further action.

US government (Republican) inquiries

State of Virginia inquiry

Mann had taught at the University of Virginia (1999–2005), during the time of
many of the CRU emails, before going to Penn State. However, the initiation of
the probe did not come from the University but from the State government. The
Republican Attorney General of Virginia, Kenneth Cuccinelli (1968–), a noted
climate skeptic, wasted no time pursuing his political agenda. Shortly after coming to office on February 17, 2010, he had attacked the EPA (*vide supra*) for using
data from the CRU in the IPCC reports:[235]

Following the close of the public comment period on the proposed
Endangerment Finding [of the EPA], internal e-mails and documents from

the Climate Research Unit at the University of East Anglia were released to the public. In these e-mails and documents, the manner in which the Climate Research Unit's global warming data was developed, analyzed, and handled was questioned. The Climate Research Unit scientists questioned the reliability of the data, and the methodologies used in developing and analyzing such data.

Attorney General Cuccinelli believes that the EPA acted in an arbitrary and capricious fashion and failed to properly exercise its judgment by relying almost exclusively on reports from the IPCC in attributing climate change to anthropogenic greenhouse gas emissions. In doing so, the EPA substantially ceded its obligation to make a judgment on the dangers of green house gas emissions. Also, the IPCC is an international body that is not subject to U.S. data quality and transparency standards and the IPCC prepared their reports in total disregard to U.S. standards ...

We cannot allow unelected bureaucrats with political agendas to use falsified data to regulate American industry and drive our economy into the ground.

Cuccinelli later accused Mann under the Virginia Fraud Against Taxpayer Act with regard to grants that the climate scientist had received from the State.[236] A Civil Investigative Demand (CID, a type of subpoena) was then issued from the Office of the Attorney General to the University of Virginia on April 27, 2010. The University of Virginia was requested to supply all records related to Mann, including any correspondence with 39 other scientists named in the CID.

The University was preparing to comply with the order but there was uproar within the faculty.[237] The scientific community at large also objected, as the blatant political interference with scientific research and academic freedom became a national and international issue. An editorial appeared in the prestigious journal, *Nature*, which stated:[238]

Certainly Cuccinelli has lost no time in burnishing his credentials with far-right "Tea Party" activists, many of whom hail him as a hero. In March, he instructed Virginia's state university presidents that they had no legal authority to protect homosexuals under their non-discrimination policies. He has also filed lawsuits challenging health-care reform and the Environmental Protection Agency's authority to issue greenhouse-gas regulations.

Cuccinelli's actions against Mann hark back to an era when tobacco companies smeared researchers as part of a sophisticated public relations strategy to raise doubts over the science showing that tobacco caused cancer, and delayed the introduction of smoking curbs for decades. Researchers found themselves bogged down in responding to subpoenas and legal challenges, which deterred others from the field. Climate-change deniers have adopted similar strategies with alacrity and, unfortunately, considerable success.

On May 27, the University filed a motion to block the investigation. The issue was finally ruled upon by the Virginia Supreme Court in favor of the University on March 2, 2012. The frivolous case had cost the State $500,000.[239] The lengthy and complex harassment of Mann by the State is summarized by the Union of Concerned Scientists.[240] Mann has since sued two groups for defamation. He stated that his reason for doing so was "the onslaught of dishonest and libelous attacks that climate scientists have endured for years by dishonest front groups seeking to discredit the case for concern over climate change."[241]

House of Representatives Select Committee on Energy Independence (ad hoc) inquiry

The Republican campaign to undermine the integrity of the IPCC and climate science extends into the highest levels of the Federal government. Allegations of scientific misconduct in the climate change sector have been made by Republicans since before the days of Kyoto.[242] An early example of the testy political commentary on climate change took place in 1995, at the House Subcommittee on Energy and the Environment.[243] And during the "eightgate" scandal, Dana Rohrabacher (1947–), Chairman of this committee, led an investigation into how Department of Energy funding was being utilized on IPCC projects.[244] On May 11, 2001, President Bush had asked the National Academy of Sciences (NAS) to determine whether the summaries of the IPCC assessments were consistent with the body of the report. The NAS concluded:[245]

> The committee finds that the full IPCC Working Group I (WG I) report is an admirable summary of research activities in climate science, and the full report is adequately summarized in the "Technical Summary". The full WG I report and its "Technical Summary" are not specifically directed at policy. The "Summary for Policymakers" reflects less emphasis on communicating the basis for uncertainty and a stronger emphasis on areas of major concern associated with human-induced climate change. This change in emphasis appears to be the result of a summary process in which scientists work with policy makers on the document. Written responses from U.S. coordinating and lead scientific authors to the committee indicate, however, that (a) no changes were made without the consent of the convening lead authors (this group represents a fraction of the lead and contributing authors) and (b) most changes that did occur lacked significant impact.

Among the members of the review committee were Jim Hansen and Richard Lindzen, who were known for their extreme and opposing views on the climate change question; other notable contributors were Robert Watson, Ben Santer, John Christy and Michael Mann.

More serious political interference began in the middle of the decade. In 2005 and 2006, Representative Joe Barton (1949–) of Texas, Chairman of the House Committee on Energy and Commerce, criticized Mann and his results.[246] On July 27, 2006, a sub-committee hearing was called by Barton on "Questions Surrounding the 'Hockey Stick' Temperature Studies: Implications for Climate

Change Assessments"; Representative Edward Whitfield (1943–) of Kentucky led the inquiry. An article on the House investigation into the hockey stick graph was later written that began with a question: "Can the U.S. Congress impeach a chart?" The answer was: yes, it can.[247]

After the climategate scandal erupted, it did not take long for it to arrive at Capitol Hill. On December 2, 2009, the US Congress Select Committee on Energy Independence and Global Warming met under the chairmanship of the noted supporter of climate legislation, Democrat Edward Markey (1946–). The meeting had been called only to provide an update of the Administration's view of the state of climate change, as the Copenhagen conference was about to begin. However, the timing of the meeting overlapped with the media frenzy over the CRU emails, and the tone of the meeting quickly degenerated as some participants became mired in the breaking story. This session was not a formal inquiry but it turned into a political inquisition on the credibility of climate science. The sides were split along party lines: the Republicans attacked the climate scientists and the Democrats dismissed the allegations.

The first member of the Committee to speak was F. James Sensenbrenner (1943–) of Wisconsin, and the discussion quickly became nasty. The Republican member immediately went to the subject of the CRU emails, and he referred to a recent commentary which stated that "the stink of intellectual corruption is overpowering"![248] The rant of the Congressman goes on with his own list of innuendoes against climate scientists. As he provided examples of the email comments from the CRU, Sensenbrenner again criticized the scientists involved and commented that the emails "read more like scientific fascism than scientific process."[249] Other Committee members joined in. John Sullivan (1965–) of Oklahoma also made inflammatory comments during the question period when he referred to the emails as "a culture of corruption in the scientific community" and that "some people … are trying to pervert the system."[250] John Shadegg (1949–) of Arizona stated:[251]

> Anyone who thinks that those e-mails are insignificant, that they don't damage the credibility of the entire movement, is naïve … These e-mails repeatedly have shown that the scientists involved and who authored them, the scientists who are behind global warming or the argument that global warming is caused by manmade factors, the e-mails demonstrate that they are afraid to reveal the facts, that they have been unwilling to have their theories tested, that they have been unwilling to provide their data, and they are unwilling to have their theories openly challenged.

As this was going on, there were only two witnesses present, who had been called upon to talk about the Administration's plans on climate change, and were not involved with the CRU emails:

1. John Holdren, Director, Office of Science and Technology Policy (President Obama's science advisor)

2.   Jane Lubchenco, Under Secretary for Oceans and Atmosphere, Administrator of the National Oceanic and Atmospheric Administration.

Neither had been asked to defend the CRU beforehand, but, under the circumstances, they had little choice. Holdren began his testimony by reassuring the Committee that the science on climate change was still valid.[252] During the tension-filled session, Holden did not dismiss the seriousness of the CRU emails; he simply suggested:[253]

> Well, I think we need to wait until all the facts are in to find out exactly what some of these e-mails mean, in terms of how the scientists in question behaved. I mean, I would point out that scientists are human, and, from time to time, they experience frustration, anger, resentment. And, from time to time, they display defensiveness and bias and even misbehavior of some kind. They are like any other group of human beings, they are subject to human frailties.

After the introductions by the two climate change experts of the Administration, the question period went immediately back to the CRU issue. Sensenbrenner started again by talking about "junk science" and "a massive international scientific fraud" and "increasing evidence of scientific fascism."[254] In an interesting reference, Sensenbrenner also mentioned "scientific McCarthyism." The Congressman was referring to how climate change scientists denounced those who do not support them.[255] The statement is ironic under the circumstances. Even after the meeting, the incensed Sensenbrenner continued his vendetta. Five days later he sent a letter, under the letterhead of the Select Committee, to Pachauri, which was offensive, stating: "The authors of the [CRU] e-mails understand what you apparently do not. Control of peer-review literature equates to control of the IPCC's conclusions."[256] He finished his letter by telling the IPCC Chairman that those involved in these emails should not be involved in IPCC activities.

The Republicans turned the CRU email scandal into a malicious crusade, even after the UK, Penn State, EPA and other inquiries, had cleared the CRU and the IPCC. During the election campaign in the fall of 2010, Representative Darrell Issa (1953–) of California stated that if the Republicans got control of the House they (the Republicans) would open another inquiry on climategate.[257]

Soon after the mid-term elections, the issue came up again at the House of Representatives. The Committee of Science and Technology, Subcommittee on Energy and Environment, held a session on November 17, 2010, with the title "A Rational Discussion on Climate Change: The Science, The Evidence, The Response." The criticism of the science of climate change began immediately. Ralph Hall (1923–) of Texas stated that the CRU emails "expose a dishonest undercurrent within the scientific ethics community."[258] But the remainder of the meeting, more or less, lived up to the title of the session. Hall was Chairman of the Science, Space and Technology Committee, where Sensenbrenner was the Vice-Chairman. On March 31, 2011, this Committee held a session on "Climate Change: Examining the Process Used to Create Science and Policy," which was to deal specifically with the issue of the CRU emails and the integrity of the IPCC.[259]

The session was a fiasco for only one of the six witnesses was not a climate skeptic. Hall mentioned a "corrupt process" with regard to the CRU emails. He has also stated elsewhere, regarding climate change, that we cannot control what God controls, and that he "was really more fearful of freezing."[260]

The new Chairman of the Science, Space and Technology Committee is another skeptic, Lamar Smith, who stated about the emails: "We now know that prominent scientists were so determined to advance the idea of human-made global warming that they worked together to hide contradictory temperature data."[261] An unusual trend emerges when looking at the Science, Space and Technology Committee. First, the old Chairman of the Committee, Hall, was a noted climate skeptic; the Vice-Chairman, Sensenbrenner, was also one; and then the new Chairman, Smith, is a climate skeptic as well. How does a radical group, even among Republicans, exercise such control over a committee in a democracy?

"Gilligangate" & the US Department of Commerce, the Inspector General inquiry

The inquisitions of the Republican Senate were just as vehement against climate change science as the House of Representatives; here is found the best known political climate skeptic – James Inhofe from Oklahoma. On July 28, 2003 Senator Inhofe, as Chair of the Committee on Environment and Public Works, stood on the Senate floor and went on his anti-climate change tirade as he gave the Senate a lecture on all the evidence against anthropogenic global warming. He began: "What I am about to do – and it is for this reason that I am doing something that is politically stupid – I am going to expose the most powerful, most highly financed lobby in Washington, the far-left environmental extremists."[262] He attacked the vague wording and uncertainty comments of the SAR.[263] And for the TAR, he criticized the SPM for being largely the work of the politicians and not the scientists.[264] His infamous "hoax" accusation appeared at the start of this long speech, and three more times near its end, including:[265]

> Let me be very clear: Alarmists are attempting to enact an agenda of energy suppression that is inconsistent with American values, freedom, prosperity, and environmental problems.
>   Over the past hour and a half I have offered compelling evidence that catastrophic global warming is a hoax. That conclusion is supported by the painstaking work of the Nation's top planet scientists.

He ended his 90-minute lament with one last mention of the "hoax" that has become his trademark:[266]

> Wake up America. With all the hysteria, all the fear, all the phony science, could it be that manmade global warming is the greatest hoax ever perpetrated on the American people? I believe it is.
>   And if we allow these detractors of everything that has made America great, those ranging from the liberal Hollywood elitists to those that are in

for the money, if we allow them to destroy the foundation, the greatest of the most highly industrialized nation in the world, then we don't deserve to live in this one nation under God. So I say to the real people: Wake up, make your voice heard. My 11 grandchildren and yours are depending on you.

Following the standard skeptic strategy, Inhofe repeated the "hoax" of climate change accusation again and again and again, regardless of the progress of the science. Another noteworthy episode took place on the floor of the US Senate in October 2004:[267]

Global warming is the greatest hoax ever perpetrated on the American people.

It was true when I said it before, and it remains true today.

Perhaps what has made this hoax so effective is that we hear over and over that the science is settled and that there is consensus that, unless we fundamentally change our way of life by limiting greenhouse gas emissions, we will cause catastrophic global warming. This is simply a false statement.

In case anyone missed his inflammatory speeches, the Senator mentioned the "hoax" in his brochure issued under the seal of the US Senate, *The Facts and Science of Climate Change* (p. 20). In this 20-page brochure, Inhofe compared global warming to the Cold War, and the IPCC to a "Soviet-style trial."[268] The title of one section of his brochure was "The IPCC Plays Hockey," a cynical reference to the work of Michael Mann, and the Senator referred to "Mann's flawed, limited research."[269] The brochure ended: "With all of the hysteria, all of the fear, all of the phony science, could it be that manmade global warming is the greatest hoax ever perpetrated on the American people? It sure sounds like it."[270]

On August 28, 2005, Inhofe made a mockery of the science by inviting noted fiction author, Michael Crichton (1942–), to be an expert witness on the science of climate change. Crichton launched a damning attack on the work of Michael Mann:[271]

Mann's work was initially criticized because his graph didn't show the well-known Medieval Warm Period, when temperatures were warmer than they are today, or the Little Ice Age, when they were colder than today. But real fireworks began when two Canadian researchers, McIntyre and McKitrick, attempted to replicate Mann's study. They found grave errors in the work, which they detailed in 2003: calculation errors, data used twice, and a computer program that generated a hockeystick out of any data fed to it – even random data. Mann's work has been dismissed as "phony" and "rubbish" by climate scientists around the world who subscribe to global warming. Some have asked why the UN accepted Mann's report so uncritically. It is unsettling to learn Mann himself was in charge of the section of the report that included his work. This episode of climate science is far from the standards of independent verification.

The inclusion of such a witness was challenged by Senators James Jeffords (1934–), the Independent from Vermont, Frank Lautenburg (1924–), the Democrat from New Jersey and Hillary Clinton (1947–), the Democrat from New York.

In February 2007 Inhofe appeared on Fox News. During the interview the Senator stated that mainstream science was beginning to accept that climate change was natural and that those who opposed this view were "those individuals on the far left, such as Hollywood liberals and the United Nations."[272] Among the more audacious actions of Senator Inhofe and other minority members of the Committee on Environment and Public Works was a report listing almost 1,000 scientists who disputed the claims about climate change.[273] Most signatories, apparently, had nothing to do with climate-related research.

On November 23, 2009, Inhofe had posted on his website that the CRU emails suggested that scientists had "cooked the science,"[274] and on December 10, he and other minority members of the Committee on Environment and Public Works sent a letter to Ban Ki-moon, the Secretary-General of the United Nations, requesting an independent (of the IPCC and UN) review which would cover:[275]

1) Allegations of adjusting or manipulating data and why various individuals refused to disclosed raw data;
2) The refusal to disclose the identity of alleged individuals or entities that have, or have been asserted to have, non-disclosure agreements with CRU and other research affiliates;
3) The refusal of various persons to release for review by other scientists the methods and actual adjustments to raw data;
4) Whether there were any attempts to influence scientific journals against publishing the work of scientists whose findings and conclusions run counter to those found in IPCC reports; and
5) Whether the IPCC's due diligence procedures governing review of IPCC assessment reports were followed and whether those procedures conform to internationally recognized peer-review norms and practices.

Another member of the Committee, Senator John Barrasso (1952–) called on Pachauri to resign; on February 4, 2010, he stood on the Senate floor and stated:[276]

Every day, new scandals emerge about the so called 'facts' in the UN reports. The integrity of the data and the integrity of the science have been compromised.

Concrete action by world leaders is needed. Government delegations of the UN's general assembly and UN Secretary Moon must pressure Dr. Rajendra Pachauri to step down as head of the United Nations Intergovernmental Panel on Climate Change.

It is time to conduct an independent investigation into the conduct of the Intergovernmental Panel on Climate Change. The scientific data behind these policies must be independently verified.

Administration policies relating to climate change will cost millions of Americans their jobs. We need to get this right. To continue to rely on these corrupted U.N. Reports is an endorsement of fraudulent behavior. It is a signal to the American people that ideology is more important than their jobs.

A Senate report from the minority members of the Committee on Environment and Public Works was then prepared on the CRU controversy and issued in February 2010. The report attacked climate science, especially the IPCC. In the introduction the Republican Senators accused climate scientists of possibly violating federal laws:[277]

After an initial review, the Minority Staff of the Senate Committee on Environment and Public Works believe the scientists involved violated fundamental ethical principles governing taxpayer-funded research and, in some cases, may have violated federal laws. Moreover, we believe the emails and accompanying documents seriously compromise the IPCC-based consensus and its central conclusion that anthropogenic emissions are inexorably leading to environmental catastrophes.

The first section of the Senate report highlighted the most damaging emails and their implications. In the second section, the allegations against the CRU were used to undermine the entire IPCC and an overview of the latter was presented.[278] Section 3, entitled "Legal and Policy Issues in the CRU Controversy," was more serious, and in it claims were made that the emails demonstrated unethical and possibly illegal behavior.[279]

Near the end of the report was a section entitled "Bios of Key Players – CRU Controversy" which listed 17 scientists, including Mann and Jones.[280] The implication of the "Group of Seventeen" is that they are the prime suspects in a covert conspiracy by the world's eminent climate scientists. Mann was public enemy number one. Inhofe was leading an inquisition against the un-American activities of eco-criminals that were threatening the United States and committing fraud with State and Federal grants, under the guise of academic freedom. Through Republican leadership the US would purge the bad science. Will the Senator soon call for trials of the "green menace"?[281]

On May 26, 2010, the zealous Senator Inhofe took the issue to a new level when he formally requested the Office of the Inspector General to investigate the CRU matter and the NOAA (National Oceanic and Atmospheric Administration). The NOAA had been dragged into this mess because it had issued a contract with the CRU back in 2003, and had previously been requested to provide more information on this matter, which it had not. Back in late November 2009, a management-level review was conducted by the NOAA. The inquiry reported back the following month that the emails raised "no concerns."[282] The Office of the Inspector General came to similar conclusions. Eight out of more than 1,000 emails had warranted further follow-up on the scientific integrity of the NOAA scientists.[283] As with all the other inquiries, no major issues or inappropriate actions were found. The one concern was the FOIA.[284]

However, within their detailed inquiry, the Office of the Inspector General may have found the "smoking gun" of the climate change conspiracy, "Gilligangate." On February 19, 2007, NOAA scientist, Thomas C. Peterson, Chief Scientist at the National Climatic Data Center, had sent an email to Jones at the CRU about the tactics of the climate skeptics:[285]

> Hi, Phil,
> I thought you might enjoy the forwarded picture and related commentary below.
> I read some of the USHCN/GISS/CRU brouhaha on web site you sent us. It is both interesting and sad. It reminds me of a talk that Fred Singer gave in which he impugned the climate record by saying he didn't know how different parts were put together. During the question part, Bob Livzey said, if you don't know how it is done you should read the papers that describe it in detail. So many of the comments on that web page could be completely addressed by pointing people to different papers. Ah well, you can lead a horse to water but you can't make it think.
>
> Warm regards,
> Tom

A picture was attached which the Office of the Inspector General commented upon.[286]

> The image depicts superimposed photographs of several individuals involved in the debate over global warming as characters from the television program Gilligan's Island, stranded on a melting ice cap at the North Pole or floating nearby in the ocean. In the course of our inquiry, we learned that another NOAA scientist had created the image during official business hours, using government computer equipment.

NOAA agreed that the image was inappropriate and those involved were "counselled."[287]

Upon completing their investigation, the Office of the Inspector General had two recommendations for the NOAA:[288]

1. Carry out a proper search for responsive records to the four FOIA requests seeking NOAA records regarding review comments on the Second Order Draft and Final Draft of the IPCC AR4 WG1, and reassess the agency's response to these requests as appropriate. Additionally, given the issues we identified in NOAA's handling of these particular FOIA requests, NOAA should consider whether these issues warrant an overall assessment of the sufficiency of its FOIA process.
2. Review the contract with CRU referenced in CRU email No. 1056478635, along with any other contracts, grants, or cooperative agreements awarded to CRU – to verify conformance with all terms and specifications, and to identify any irregularities – and provide the results of its review to us.

Undeterred, Inhofe has continued his campaign; in 2012

- his book, *The Greatest Hoax: How the Global Warming Conspiracy Threatens Your Future*, was published
- he proposed to slash the budget of the "rogue" EPA[289]
- He stated: "My point is, God's still up there ... The arrogance of people to think that we, human beings, would be able to change what He is doing in the climate is to me outrageous."[290]

The Republican Senators appear to have had no restraints imposed upon them on how they could wield their legislative powers, and they freely intimidated, threatened and insulted some of the most famous climate scholars in the world. While morally corrupt, the tactic has worked:[291]

> Conservatives attempted to intimidate these high-profile scientists, and sully their image, with the aim of disparaging mainstream climate science more generally by association. While their tactics produced no evidence of scientific wrongdoing, their public accusations nevertheless were enough to cast doubt on climate science within Congress.

Climate scientists have been put on the defensive: "You can't have a spelling mistake in a paper without it being evidence on the floor of the Senate that the system is corrupt."[292]

In an issue of *Nature*, there was a commentary called "Climate of Fear"; the subtitle read: "The integrity of climate research has taken a very public battering in recent months. Scientists must now emphasize the science, while acknowledging that they are in a street fight."[293] One climate scientist stated in the same article: "Everyone is scared shitless, but they don't know what to do."[294]

The US National Academy of Sciences was outraged by what was taking place; on May 7, 2010, a letter signed by many members appeared in the journal *Science*:[295]

> WE ARE DEEPLY DISTURBED BY THE RECENT ESCALATION OF POLITICAL ASSAULTS ON SCIENTISTS in general and on climate scientists in particular ...
>
> Many recent assaults on climate science and, more disturbingly, on climate scientists by climate change deniers, are typically driven by special interests or dogma, not by an honest effort to provide an alternative theory that credibly satisfies the evidence ...
>
> We also call for an end to McCarthy-like threats of criminal prosecution against our colleagues based on innuendo and guilt by association, the harassment of scientists by politicians seeking distractions to avoid taking action, and the outright lies being spread about them. Society has two choices: We can ignore the science and hide our heads in the sand and hope we are lucky,

or we can act in the public interest to reduce the threat of global climate change quickly and substantively. The good news is that smart and effective actions are possible. But delay must not be an option.

A recent study has found that skepticism about climate change is largely an Anglo-Saxon phenomenon, especially in the US and UK, and is absent in Continental Europe and other nations.[296] The great climate schism in America is split along party lines.[297] And there is little evidence that the Republican skeptics can be swayed from their views by more science:[298]

> The moderating effect of political orientation challenges the common assumption of climate change communicators (e.g., scientists and policy-makers) that more information or education will help convince Americans of the need to deal with climate change. Particularly for those on the Right, this seems unlikely to prove effective. Our results, along with those of prior studies, show that education and self-reported understanding of global warming have little effect on the views of climate change held by Republicans and conservatives. Reducing climate skepticism among this large segment of the American populace will require far more than simply providing additional information.
>
> Our results indicate that this conflict is also diffusing throughout the American public. Liberals and Democrats are more likely to take the side of the scientific consensus and many environmental movement organizations, proclaiming that global warming is real, is human-caused, and is a worrisome threat. On the other hand, conservatives and Republicans are more likely to dispute or deny the scientific consensus and the claims of the environmental community, thereby defending the industrial capitalist system.

Similar conclusions were raised in another study:[299]

> The more important factor at the aggregate level is the polarized positions taken by Democrats and Republicans. This polarization over environmental issues is long standing, and has extended to the climate-change arena, making it a highly partisan issue. Given the vested economic interests reflected in this polarization, it seems doubtful that any communication process focused on persuading individuals will have much impact ... Political conflicts are ultimately resolved through political mobilization and activism. Further efforts to address the issue of climate change need to take this into account.

An ironic aspect of the Republican hate campaign is their direct connections to the IPCC and the AR4. The IPCC had been created because of the initiative of a Republican President (Ronald Reagan). During the first decade of the millennium, when the CRU emails were sent, most American members of the IPCC had been appointed by another Republican President (George W. Bush) and Pachauri had been supported by the Bush Administration. The Republican appointees within the IPCC had major influence over the preparation of the AR4. In addition, it was under the Bush Administration that the American government officially accepted and approved the contents of the AR4. One is astounded, then, that the IPCC and AR4 would be so vilified by many Republicans since, in many ways, it is *their* document.

The skeptic movement

Known as "climate change skeptics," or simply "climate skeptics," these deniers of climate change have dedicated themselves to a simple task – stop any climate change legislation. The opposition is not new as reported by the frustrated Chair of the IPCC: "[H]onest scientific discourse wilts under gross distortions and ideo-logically driven posturing. Sadly, such tactics have been a prominent feature of climate science for many years."[300]

Back in 1991, the noted climatologist, Stephen Schneider, warned that:[301]

> excessive claims and counterclaims were played up in the media. The media circus spread confusion and doubt among citizens and policymakers alike – at least in the United States – to the point that ... any resolve to slow global warming was in political trouble.

He later added: "[C]limate scientists willing to enter the fray would have to join a rough-and-tumble free for all – no rules and no refs in this contact sport."[302]

By the late 1980s a few scientists had made a name for themselves as climate skeptics, including Patrick J. Michaels (1950–), Richard Lindzen (1940–)[303] and Siegfried Fred Singer (1924–).[304] A more organized contrarian movement in the US appeared shortly after the FAR had been issued by the IPCC.[305] Behind many of the leading climate skeptics is a symbiotic relationship with media-savvy, con-servative think tanks or corporations, which provide financial support for them as fellows, visiting scholars and consultants.[306] The major organized support groups are:

- fossil fuel and related corporate entities and trade associations
- right-wing think tanks
- ultra-conservative politicians.

Among the early organizations of the climate change skeptic movement[307] had been the:

- Global Climate Coalition (GCC)[308] – Don Pearlman, who was also associated with the Global Climate Council, John Schlaes[309] and William O'Keefe, who was later associated with the Marshall Institute
- George C. Marshall Institute (Marshall Institute) – William Nierenberg (1919–2000) [310] and Fredrick Seitz, who was also associated with the SEPP of Fred Singer.

A tactic of the skeptics is to discredit the Assessment Reports by identifying hon-est, taken out of context or deliberately misinterpreted "mistakes." The propa-ganda is masterfully orchestrated by repeating over and over again that legislation to restrict greenhouse gas emissions would undermine economic growth and cost jobs – powerful sound bites that catch the attention of the general public:[311]

> Several conservative think tanks, political leaders, and commentators continue to hew closely to their decade-old playbook for downplaying the urgency of climate change, which includes questioning whether human activities drive climate change while also arguing that any action to curb it will lead to dire economic consequences.

The basic assault weapons of the climate skeptics are propaganda of fearmonger-ing and doubt under the guise of sound science, played out through social and public media outlets. Blogs are the skeptics' journals, where fear review replaces peer review. Back in 1994, Bolin had hoped that the peer-review process would be a barrier to the attack of skeptics:[312]

> Instead, this controversy has rather been presented in special reports and the popular press. Key scientific issues cannot be resolved in this way. Scientific analyses should be submitted to the relevant scientific journals so that they can be thoroughly scrutinized in the assessment process.

Bolin's vision did not come true, as much of the climate change debate took place outside of recognized scientific circles:[313]

> One of the most obvious features of the climate change debate is the influence of the blogosphere. This provides an opportunity for unmoderated comment to stand alongside peer reviewed publications; for presentations or lectures at learned conferences to be challenged without inhibition; and for highly per-sonalized critiques of individuals and their work to be promulgated without hindrance.

Whereas scientific papers can take years of research and months to get accepted in a peer-reviewed journal, blogs can be churned out in a few minutes and are only limited by one's imagination and desire to produce them. In the anarchistic world of the blogosphere, a website can become a stage in the global theatre, from which an actor posing as an expert or a mercenary scientist can influence the public more than a stuffy peer-reviewed article in a scientific journal. The scientific media is overwhelmed by social media.[314]

Skepticism, of course, is a necessary part of scientific thinking. One is reminded of the great scientist Robert Boyle (1627–91), whose famous work was called the *Skeptical Chymist*. In his pioneering study, he believed in skepticism, but where experiments provided support for the theories of science; whereas, this "modern" skepticism relies on rhetoric and belongs to an era that predates Boyle, the Dark

Ages, when dogma determined the validity of the theories of science. Not since the days of Copernicus (1473–1543) and Galileo (1564–1642), when scientists dared to claim that the Earth revolved around the sun, has a scientific theory caused so much outrage. In the past, stubborn naysayers and rulers rejected the new scientific truths, and some scientists faced the Inquisition. With climate change, the story unfolds in much the same way as politics again challenges science, especially in America. Sadly, the brunt of these allegations has fallen on those scholars associated with the IPCC, men and women who have freely offered their time, responding to a call by the United Nations to help all of humanity. For their trouble, they have been harassed and slandered by a small, but vocal, opposition. The real problem, though, is that certain politicians have even tried to raise criminal charges against them. Does it seem right that American citizens, many internationally recognized for their contributions to science, are bullied by elected officials? At this political level, the skeptics' movement is a travesty of justice, not only against the scientific community but the rights of citizens in general, that is unprecedented in a democracy.

References[315]

Absai, R., UN Climate Chief Rules Out Conspiracy Behind Debate, *Gulf News*, December 13, 2009; http://gulfnews.com/news/gulf/uae/environment/un-climate-chief-rules-out-conspiracy-behind-debate-1.552662

Adam, D., Climate Scientist Says Himalayan Glacier Report is 'Robust and Rigorous,' *Guardian*, February 14, 2010; http://www.guardian.co.uk/environment/2010/feb/14/climate-scientist-himalayan-glacier-report?INTCMP=SRCH

Adam, D., Climate: The Hottest Year, *Nature* 468, 362, November 15, 2010

Alleyne, R., UN Report on Glacier Melting is Based on Speculation, *The Telegraph*, January 17, 2010; http://www.telegraph.co.uk/earth/earthnews/7011713/UN-report-on-glaciers-melting-is-based-on-speculation.html

Anon., Climate Debate Must Not Overheat, *Nature* 381, June 13, 1996

Anon., Climate of Fear, *Nature* 464, 141, (March 11), 2010

Anon., Science Subpoenaed, *Nature* 465, 135, (May 13), 2010

Antilla, L., Climate of Skepticism: U.S. Newspaper Coverage of the Science of Climate Change, *Global Environmental Change* 15, 338, 2005

Arthur, C., Hacking into the Mind of the CRU Climate Change Hacker, *Guardian*, February 5, 2010; http://www.guardian.co.uk/environment/blog/2010/feb/05/cru-climate-change-hacker

Assmann, S.M., Castleman, W., Irwin, M.J., Jablonski, N.G., and Vondracek, F.W., *RA-10 Final Investigation Report Involving Dr. Michael M. Mann*, Pennsylvania State University, June 4, 2010; http://live.psu.edu/pdf/Final_Investigation_Report.pdf; accessed November 2011

Bagla, P., Climate Science Leader Rajendra Pachauri Confronts the Critics, *Science* 327, 510, January 29, 2010

Bagla, P., Himalayan Glaciers Bucking the Trend (IPCC's Himalayan Blunder), New Delhi Television, November 9, 2009; http://www.youtube.com/watch?v=bnYmQjFoNCs

Bagla, P., Himalayan Glaciers Melting Deadline 'A Mistake', BBC News, December 5, 2009; http://news.bbc.co.uk/2/hi/south_asia/8387737.stm

Bagla, P., No Sign Yet of Himalayan Meltdown, Indian Report Finds, *Science* 326, 924, November 13, 2009

Banerjee, B., and Collins, G., Anatomy of IPCC's Mistake on Himalayan Glaciers and 2035, *The Yale Forum on Climate Change & the Media*, February 4, 2010; http://www. yaleclimatemediaforum.org/2010/02/anatomy-of-ipccs-himalayan-glacier-year-2035-mess/

Barrasso, J.B., Barrasso Calls on UN Climate Chief's Resignation, Press Office, February 4, 2010; http://barrasso.senate.gov/public/index.cfm?FuseAction=PressOffice.Press Releases&ContentRecord_id=9A9B602C-D104–2C0E-E6F1–355602F94FE2

BBC News, Global Warming Biggest Threat, January 9, 2004; http://news.bbc.co.uk/2/ hi/3381425.stm; accessed October 2012

BBC News, Hackers Target Leading Climate Research Unit, November 20, 2009; http:// news.bbc.co.uk/2/hi/science/nature/8370282.stm

Bolin, B., *A History of the Science and Politics of Climate Change*, Cambridge University Press, 2007

Bolin, B., Next Step for Climate-change Analysis, *Nature* 368, 94, 1994

Bolin, B., Science and Policymaking, *Ambio* 23 (1), 25, 1994

Booker, C., Pachauri: The Real Story Behind the Glaciergate Scandal, *The Telegraph*, January 23, 2010; http://www.telegraph.co.uk/comment/columnists/christopherbooker/7062667/ Pachauri-the-real-story-behind-the-Glaciergate-scandal.html

Brainard, C., Q&A: Covering the IPCC, *The Observatory*, September 14, 2010; http:// www.cjr.org/the_observatory/qa_covering_the_ipcc.php

Bravender, R., Rep. Issa Would Lead 'Climategate' Probe if House Goes to GOP, *New York Times*, September 23, 2010; http://www.nytimes.com/cwire/2010/09/23/23climatewire-rep-issa-would-lead-climategate-probe-if-hou-44766.html?scp=4&sq=%22climatic%20 research%20unit%22&st=cse

Brulle, R.J., Carmichael, J., and Jenkins, J.C., Shifting Public Opinion on Climate Change, *Climatic Change*, February 2, 2012

Christy, J.R., IPCC: Cherish It, Tweak It, or Scrap It, *Nature*, 463, 732, February 11, 2010

Cuccinelli, K.T., *Cuccinelli Petitions EPA and Files for Judicial Review*, February 17, 2010; http://www.oag.state.va.us/PRESS_RELEASES/Cuccinelli/21710_Attorney_ General%20Petitions%20EPA.html; accessed November 2011

Dawson, B., Analyzing Headlines and Lead Sentences in Penn State's Michael Mann Inquiry, The Yale Forum on Climate Change & the Media, March 18, 2010; http://www. yaleclimatemediaforum.org/2010/03/headlines-and-lead-sentences/

Delingpole, J., Climategate: the Final Nail in the Coffin of 'Anthropogenic Global Warming'?, *The Telegraph*, November 20, 2009; http://blogs.telegraph.co.uk/news/jamesdelingpole/ 100017393/climategate-the-final-nail-in-the-coffin-of-anthropogenic-global-warming/

Down to Earth, Glaciers Beating Retreat, April 30, 1999; http://www.downtoearth.org.in/ content/glaciers-beating-retreat

Earth Negotiations Bulletin; http://www.iisd.ca/vol12/

Edwards, P.N., and Schneider, S.H., Broad Consensus or "Scientific Cleansing," *Ecofable/ Ecoscience* 1 (1), p. 3, 1997; http://pne.people.si.umich.edu/PDF/ecofables.pdf

EPA, *Denial of Petitions for Reconsideration of the Endangerment and Cause or Contribute Findings for Greenhouse Gases under Section 202(a) of the Clean Air Act*, last updated 9 September 2013, see *Response to Petitions*; http://epa.gov/climatechange/endangerment/ petitions.html

Federal Register, 75 (156), p. 49557, August 13, 2010; https://www.federalregister. gov/articles/2010/08/13/2010-19153/epas-denial-of-the-petitions-to-reconsider-the-endangerment-and-cause-or-contribute-findings-for

Financial Post, Open Climate Letter to UN Secretary-General, November 30, 2012; http://opinion.financialpost.com/2012/11/29/open-climate-letter-to-un-secretary-general-current-scientific-knowledge-does-not-substantiate-ban-ki-moon-assertions-on-weather-and-climate-say-125-scientists/

Foley, H.C., Scaroni, A.W., and Yekel, C.A., *RA-10 Inquiry Report: Concerning the Allegations of Research Misconduct Against Dr. Michael E. Mann, Department of Meteorology, College of Earth and Mineral Sciences, The Pennsylvania State University*, February 3, 2010; http://www.research.psu.edu/orp/Findings_Mann_Inquiry.pdf

Gelbspan, R., *The Heat is On: The Climate Crisis, The Cover-up, The Prescription*, Basic Books, 1997

Girling, R., The Leak Was Bad. Then Came the Death Threats, *The Sunday Times*, February 7, 2010; http://www.timesonline.co.uk/tol/news/environment/article7017905.ece; accessed November 2011

Gleick, P.H., *et al.*, Climate Change and the Integrity of Science, *Science* 328, 689, May 7, 2010

Goldenberg, S., Obama Environment Agenda Under Threat from Incoming Republicans, *Guardian*, October 31, 2010; http://www.guardian.co.uk/world/2010/oct/31/republican-onslaught-obama-environment-agenda

Hasnain, S.I., *Report on Himalayan Glaciology*, 1999; supplied by an email attachment from Dr. Hasnain, sent on July 1, 2011

Heiderman, R.S., U-Va Urged to Fight Subpoena of Climate Scientist's Documents, *Washington Post*, May 9, 2010; http://www.washingtonpost.com/wp-dyn/content/article/2010/05/08/AR2010050802020.html

Hickman, L., and Randeson, J., Climate Skeptics Claim Leaked eMails are Evidence of Collusion among Scientists, *Guardian*, November 20, 2009; http://www.guardian.co.uk/environment/2009/nov/20/climate-sceptics-hackers-leaked-emails

Hindustantimes, Can't Totally Depend on IPCC, India to have Own Climate Panel, February 4, 2010; http://www.hindustantimes.com/India-to-have-own-panel-on-climate-change-Jairam-Ramesh/H1-Article1-505242.aspx

Hulme, M., IPCC: Cherish It, Tweak It, or Scrap It, *Nature*, 463, 730, February 11, 2010

Hulme, M., The IPCC's Problems Have Been Compounded by its Impervious Attitude, *Guardian*, February 5, 2010; http://www.guardian.co.uk/environment/2010/feb/05/rajendra-pachauri-hacked-climate-science-emails?INTCMP=SRCH

Hulme, M., and Ravitz, J., 'Show Your Working': What ClimateGate Means, BBC News, December 1, 2009; http://news.bbc.co.uk/2/hi/8388485.stm

Inhofe, J., *Inhofe Delivers Major Speech on the Science of Climate Change*, July 29, 2003; http://www.inhofe.senate.gov/epw-archive/press/bsen-inhofe-delivers-major-speech-on-the-science-of-climate-change/b-icatastrophic-global-warming-alarmism-not-based-on-objective-sciencei-ipart-1/i

Inhofe, J., *The Facts and Science of Climate Change*, n.d.; http://epw.senate.gov/public/index.cfm?FuseAction=Files.View&FileStore_id=01d83873-cb56-4153-9b8d-f9dd65366b0c

Inhofe, J., *The Greatest Hoax: How the Global Warming Conspiracy Threatens Your Future*, WND Books, 2012

Inhofe, J., Science of Climate Change, *Congressional Record*, S10012, July 28, 2003; http://www.gpo.gov/fdsys/pkg/CREC-2003-07-28/pdf/CREC-2003-07-28-pt1-PgS10012.pdf

InterAcademy Council, *Climate Change Assessments, Review of the Processes and Procedures of the IPCC*, 2010; http://www.interacademycouncil.net/24026/26050.aspx

InterAcademy Council, *Lighting the Way: Toward a Sustainable Energy Future*, 2007; http://www.interacademycouncil.net/24770/25271.aspx

IPCC, *Fourth Assessment Report, Working Group II: Impacts, Adaptations and Vulnerability*, 2007

IPCC, *Fourth Assessment Report, Working Group II: Impacts, Adaptations and Vulnerability, Errata*, 2007

IPCC, *Fourth Assessment Report, Working Group II: Impacts, Adaptations and Vulnerability, Government and Expert Review of Second Order Draft, Specific Comments,*[317] *Expert Review Comments, Chapter 10*, August 2006

IPCC, *Fourth Assessment Report, Working Group II: Impacts, Adaptations and Vulnerability, Government and Expert Review of Second Order Draft, Specific Comments, Expert Review Comments, Executive Summary*, August 2006

IPCC, *Fourth Assessment Report, Working Group II: Impacts, Adaptations and Vulnerability, Government and Expert Review of Second Order Draft, Specific Comments, Government Review Comments, Chapter 10*, August 2006

IPCC, *Fourth Assessment Report, Working Group II: Impacts, Adaptations and Vulnerability, Government and Expert Review of Second Order Draft, Specific Comments, Government Review Comments, Summary for Policymakers*, August 2006

IPCC, *IPCC Statement on News Reports Regarding Hacking of the East Anglia University Email Communications*, Statements,[318] December 4, 2009; http://www.ipcc.ch/pdf/presentations/rkp-statement-4dec09.pdf

IPCC, *IPCC Statement on the Melting of the Himalayan Glaciers*, Statements, January 20, 2010

IPCC, *Launch of Independent Review of the IPCC Processes and Procedures*, Press Releases,[319] March 10, 2010

IPCC, *Opening Statement by Dr. Rajendra Pachauri, Chairman of the Intergovernmental Panel on Climate Change at a Press Conference at the United Nations in New York*, Statements, August 30, 2010

IPCC, *Report of the Eleventh Session of the Intergovernmental Panel on Climate Change*,[320] December 11–15, 1995

IPCC, *Report of the Thirteenth Session of the Intergovernmental Panel on Climate Change*, Maldives, September 22 & 25–28, 1997

IPCC, *Report of the Seventeenth Session of the Intergovernmental Panel on Climate Change, Future Work Programme of the IPCC*, April 2001

IPCC, *Report of the Nineteenth Session of the Intergovernmental Panel on Climate Change, Evaluation of the Work of Working Group III and Its Technical Support Unit from 1998–2002: Lessons for the Future*, April 2002

IPCC, *Report of the 28th Session of the IPCC Bureau*, Geneva, December 10–11, 2002; http://www.ipcc.ch/meeting_documentation/meeting_documentation_sessions_of_the_ipcc_bureau.shtml

IPCC, *Report of the 29th Session of the IPCC*, Geneva, August 31 – September 4, 2008

IPCC, *Report of the 29th Session of the IPCC, Address by Ban Ki-moon*, September 2008

IPCC, *Report of the 30th Session of the IPCC, Future IPCC Activities, Reinforcement of the IPCC Secretariat – Report from the Task Group*, April 2009

IPCC, *Report of the 31st Session of the IPCC*, Bali, Indonesia, October 26–29, 2009

IPCC, *Draft Report of the 32nd Session of the IPCC*,[316] Busan, Republic of Korea, October 11–14, 2010

IPCC, *Report of the 33rd Session of the IPCC, Review of the IPCC Processes and Procedures, Draft Proposals by the Task Group on Governance and Management, Terms of Reference for the Bureau and Bureau Members*, May 10–13, 2011

IPCC, *Report of the 33rd Session of the IPCC, Decisions Taken with Respect to the Review of IPCC Processes and Procedures, Governance and Management*, May 10–13, 2011

IPCC, *Report of the 33rd Session of the IPCC, Decisions Taken with Respect to the Review of IPCC Processes and Procedures, Procedures*, May 10–13, 2011

IPCC, *Report of the 33rd Session of the IPCC, Review of the IPCC Processes and Procedures, Proposal by the Task Group on Procedures*, May 10–13, 2011

IPCC, *Report of the 34th Session of the IPCC, Further Work to Adopting Revisions to "Appendix C of the Principles Governing IPCC Work: Rules of Procedures for the Election the of IPCC Bureau and Any Task Force Bureau,"* November 18–19, 2011

IPCC, *Report of the 34th Session of the IPCC, Review of IPCC Processes and Procedures, Procedures: Adoption of the revised "Appendix A to the Principles Governing IPCC Work: Procedures for the Preparation, Review, Acceptance, Adoption, Approval and Publications of IPCC Reports" Role of Review Editors – Revised Guidance Notes*, November 18–19, 2011

IPCC, *Review by Dutch Government Confirms IPCC Core Conclusions on Impacts of Climate Change – Recommendations for Future Improvements Welcomed*, Press Release, July 5, 2010

IPCC, *Second Assessment Report, Working Group I: The Science of Climate Change*, 1995

IPCC, *Statement of the IPCC Chairman on the Establishment of an Independent Committee to Review IPCC Procedures*, Geneva, Statements, February 27, 2010

IPCC, *Statement on News Reports Regarding Hacking of the East Anglia University EMail Communications*, Statements, December 4, 2009

IPCC, *Third Assessment Report, Working Group I: The Scientific Basis*, 2001

Jarvis, B., The 10 Dumbest Things Ever Said About Global Warming, *Rolling Stone*, June 19, 2013; http://www.rollingstone.com/politics/news/the-10-dumbest-things-ever-said-about-global-warming-20130619

Jha, A., Freedom of Information Laws are used to Harass Scientists, says Nobel Laureate, *Guardian*, May 25, 2011; http://www.guardian.co.uk/politics/2011/may/25/freedom-information-laws-harass-scientists?INTCMP=SRCH

Kintisch, E., Stolen E-mails Turn Up Heat on Climate Change Rhetoric, *Science*, 326, 1329, December 4, 2009

Lacey, S., Rep. Lamar Smith, Who Criticized 'the Idea of Human-made Global Warming,' Set to Chair House Science Panel, *Thinkprogress, Climate Progress*, November 28, 2012; http://thinkprogress.org/climate/2012/11/28/1248751/rep-lamar-smith-who-criticized-the-idea-of-human-made-global-warming-set-to-be-chair-of-house-science-panel/

Lacey, S., Senator Inhofe and the Heartland Institute Roll Out Underwhelming Campaign to Slash the EPA, *Climate Progress*, November 27, 2012; http://thinkprogress.org/climate/2012/11/27/1241161/senator-inhofe-and-the-heartland-institute-roll-out-underwhelming-campaign-to-slash-the-epa/

Lahsen, M. (Marcus, G.E., ed.), *The Detection and Attribution of Conspiracies: the Controversy over Chapter 8, Paranoia Within Reason: A Casebook on Conspiracy as Explanation*, 111, University of Chicago Press, 1999

Leake, J., and Hastings, C., World Misled Over Himalayan Glacier Meltdown, The Heartland Institute, January 17, 2010; http://heartland.org/policy-documents/world-misled-over-himalayan-glacier-meltdown

Lean, G., IPCC Chief Rajendra Pachauri to Face Independent Inquiry, *Daily Telegraph*, February 26, 2010; http://www.telegraph.co.uk/earth/environment/climatechange/7316758/IPCC-chief-Rajendra-Pachauri-to-face-independent-inquiry.html

Lean, G., Rajendra Pachauri to Defend Handling of IPCC After Climate Change Science Row, *Daily Telegraph*, February 24, 2010; http://www.telegraph.co.uk/earth/environment/climatechange/7308210/Rajendra-Pachauri-to-defend-handling-of-IPCC-after-climate-change-science-row.html;

Leemans, R., Personal Experiences with the Governance of the Policy-Relevant IPCC and Millennium Ecosystems Assessment, *Global Environmental Change* 18, 12, 2008

Leggett, J., Carbon Wars, *Guardian*, April 25, 2006; http://www.guardian.co.uk/commentisfree/2006/apr/25/exxonmobilslonglivedemulatio

Leiserowitz, A., Mailbach, E., Roser-Renouf, C., and Smith, N., Global Warmings Six Americas, *Yale Project on Climate Change Communications*, May 2011; http://environment.yale.edu/climate/publications/SixAmericasMay2011

Maassarani, T., Redacting the Science of Climate Change, 2007; http://www.whistleblower.org/storage/documents/RedactingtheScienceofClimateChange.pdf

Mann, M., Bradley, R.S., and Hughes, M.K., Northern Hemisphere Temperatures in the Past Millennium: Interferences, Uncertainties, and Limitations, *Geophysical Research Letters*, 26 (6), p. 759, 1999; also see Mann, M.E., Bradley, R.S., and Hughes, M.K., *Global-scale Temperature Patterns and Climate Forcing Over the Past Six Centuries*, *Nature* 392, 779, 1998

Masood, E., Climate Report Subject to Scientific Cleansing, *Nature* 381 (6583), 546, June 13, 1996

May, B., Under-Informed, Over Here, *Guardian*, January 27, 2005; http://www.guardian.co.uk/science/2005/jan/27/lastword.environment

McCarthy, M., Glaciergate Part II: Climate Chief Reheats Himalayas Row, *Independent*, December 7, 2011; http://www.independent.co.uk/environment/climate-change/glaciergate-part-ii-climate-chief-reheats-himalayas-row-6273288.html

McCrae, F., Professor in Climate Change Scandal Helps Police with Enquiries while Researchers Call for him to be Banned, *Mail Online*, December 2, 2009; http://www.dailymail.co.uk/news/article-1232722/Professor-climate-change-scandal-helps-police-enquiries-researchers-banned.html

McCright, A.M., and Dunlap, R.E., Anti-reflexivity: The American Conservative Movement's Success in Undermining Climate Science and Policy, *Theory, Culture & Society* 27 (2–3), 100, 2010

McCright, A.M., and Dunlap, R.E., Defeating Kyoto: the Conservative Movement's Impact on U.S. Climate Change Policy, *Social Problems* 50 (3), 348, 2003

McCright, A.M., and Dunlap, R.E., The Politicization of Climate Change and Polarization in the American Public's View of Global Warming, 2001–2010, *Sociological Quarterly* 52 (2), 155, 2011

McCullagh, D., Congress May Probe Global Warming Leaked E-Mails, CBS News, November 24, 2009; http://www.cbsnews.com/8301-504383_162-5761180-504383.html

McKitrick, R., *Understanding the Climategate Inquiries*, September 2010; http://rossmckitrick.weebly.com/climategate.html

Monastersky, R., Climate Science on Trial, *Chronicle of Higher Education*, 53 (3), 14, 2006

Montford, A., *The Climategate Inquiries*, The Global Warming Policy Foundation, September 2010; http://www.thegwpf.org/gwpf-reports/1531-the-climategate-inquries.html

Morano, M., Geologist 'appalled' at NYT's Krugman: 'Legitimate scientists do not doctor data ... hijack peer-review ... send fraudulent data to UN that is used to perpetuate greatest hoax in the history of science', *Climate Depot*, November 29, 2009; http://www.climatedepot.com/a/4140/Geologist-appalled-at-NYTs-Krugman-Legitimate-scientists-do-not-doctor-datahijack-peerreviewsend-fraudulent-

National Research Council, National Academy of Sciences, *Climate Change Science: An Analysis of Some Key Questions*, The National Academies Press, 2001; http://www.nap.edu/openbook.php?record_id=10139&page=R7

National Science Foundation, *Closeout Memorandum A09120086*, August 15, 2011; http://www.nsf.gov/oig/search/results.cfm; accessed February 2012

Netherlands Environmental Assessment Agency, *Assessing an IPCC Assessment: An Analysis of Statements on Projected Regional Impacts in the 2007 Report*, 2010; http://www.pbl.nl/en/publications/2010/Assessing-an-IPCC-assessment.-An-analysis-of-statements-on-projected-regional-impacts-in-the-2007-report

Nisbet, M.C., Communicating Climate Change, *Environment* 51 (2), 14, 2009

Oreskes, N., and Conway, E., Climate Change Denial: A History, *New Statesman*, June 2010; http://www.newstatesman.com/global-issues/2010/05/climate-scientists-science

Oxburgh, R., Davies, H., Emanuel, K., Graumlich, L., Hand, D., Huppert, H., and Kelly, M., *Report of the International Panel Set Up by the University of East Anglia to Examine the Research of the Climatic Research Unit*, April 19, 2010; http://www.uea.ac.uk/mac/comm/media/press/CRUstatements/oxburgh

Pachauri, R., *IAC Review of the IPCC*, Committee Meeting 1 – Public Session, May 14, 2010; http://reviewipcc.interacademycouncil.net/

Painter, J., *Poles Apart, The International Reporting of Climate Scepticism*, University of Oxford, 2011

Painter, J., and Ashe, T., Cross-National Comparison of Climate Scepticism in the Print Media in Six Countries, 2007–10, *Environmental Research Letters*, October 4, 2012

Patwardhan, A., *Brief Review of the IPCC Process*, SREX, Presentations, January 7, 2010; http://www.adpc.net/IPCC/presentations/default.htm

Pearce, F., Debate Heats Up Over IPCC Melting Glacier Claim, *New Scientist*, January 11, 2010; http://www.newscientist.com/article/dn18363-debate-heats-up-over-ipcc-melting-glaciers-claim.html

Pearce, F., Flooded Out, *New Scientist*, 162 (2189), 18, June 5, 1999

Pearce, F., Frankenstein Syndrome Hits Climate Treaty, *New Scientist*, 142 (1929), 5, June 11, 1994

Pielke Sr., R., *Comment on the Post "Enemies Caught in Action!" On the Blackboard*, November 22, 2009; http://wattsupwiththat.com/2009/11/22/pielke-senior-comment-on-the-post-%E2%80%9Cenemies-caught-in-action%E2%80%9D-on-the-blackboard/

Piltz, R., IPCC Slips on the Ice with Statement about Himalayan Glaciers, *Climate Science Watch*, January 19, 2010; http://www.climatesciencewatch.org/2010/01/19/ipcc-slips-on-the-ice-with-statement-about-himalayan-glaciers/

Price, J., IPCC: Cherish It, Tweak It, or Scrap It, *Nature* 463, 732, February 11, 2010

Raina, V.K., *Himalayan Glaciers: A State-of-the-Art Review of Glacier Studies, Glacier Retreat and Climate Change*, Ministry of Environment and Forests, November 12, 2009; http://scienceandpublicpolicy.org/reprint/himalayan_review_of_glacial_studies.html

Ramesh, R., India 'Arrogant' to Deny Global Warming Link to Melting Glacier, *Guardian*, November 9, 2009; http://www.guardian.co.uk/environment/2009/nov/09/india-pachauri-climate-glaciers?INTCMP=SRCH

Ravilious, K., Hacked eMail Climate Scientists Receive Death Threats, *Guardian*, December 8, 2009; http://www.guardian.co.uk/environment/2009/dec/08/hacked-climate-emails-death-threats?INTCMP=SRCH

Ravindranath, N.H., IPCC: Accomplishments, Controversies and Challenges, *Current Science* (Bangalore), 99 (1), 26, 2010

Revkin, A.C., A Climate Scientist Who Engages Skeptics, *New York Times*, November 27, 2009; http://dotearth.blogs.nytimes.com/2009/11/27/a-climate-scientist-on-climate-skeptics/#more-11377

Revkin, A.C., Hacked E-Mail Is New Fodder for Climate Dispute, *New York Times*, p. A1, November 21, 2009; http://www.nytimes.com/2009/11/21/science/earth/21climate.html?_r=4

Revkin, A.C., Nobel Halo Fades Fast for Climate Change Panel, *New York Times*, August 3, 2009; http://www.nytimes.com/2009/08/04/science/earth/04clima.html?ref=intergovernmentalpanelonclimatechange

Revkin, A.C., Politics Reasserts Itself in the Debate Over Climate Change and its Hazards, *New York Times*, August 5, 2003

Revkin, A.C., The East Anglia Climate Snatch – Not (Yet) a Crime, *New York Times*, March 1, 2011; http://dotearth.blogs.nytimes.com/2011/03/01/the-east-anglia-climate-snatch-not-yet-a-crime/?scp=5&sq=%22climatic%20research%20unit%22&st=cse

Rose, D., Glacier Scientist: I Knew Data Hadn't Been Verified, *Mail Online*, January 24, 2010; http://www.dailymail.co.uk/news/article-1245636/Glacier-scientists-says-knew-data-verified.html

Russell, M., Boulton, G., Clarke, P., Eyton, D., and Norton, J., *The Independent Climate Change eMail Review*, July 2010; http://www.cce-review.org/pdf/FINAL%20REPORT.pdf

Russell, W.G., Civil Investigative Demand, April 27, 2010; http://voices.washingtonpost.com/virginiapolitics/Virginia_Attorney_General_CID.pdf

Santer, B., *et al.*, Response to Wall Street Journal Editorial of June 12th, 1996 by Frederick Seitz, *Wall Street Journal*, June 25, 1996

Schneider, S., *Science as a Contact Sport, Inside the Battle to Save Earth's Climate*, National Geographical Society, 2009

Schneider, S., Three Reports of the Intergovernmental Panel on Climate Change, *Environment*, 33 (1), 25, 1991

Science Daily, Climate Change Beliefs: Political Views Trump Facts for Some, October 4, 2012; http://www.sciencedaily.com/releases/2012/10/121004134731.htm?utm_source=feedburner&utm_medium=email&utm_campaign=Feed%3A+sciencedaily%2Ftop_news%2Ftop_environment+%28ScienceDaily%3A+Top+News+--+Top+Environment%29; accessed October 2012

Seitz, F., Major Deception of Global Warming, *Wall Street Journal* 227 (115), A16, June 12, 1996

Sensenbrenner, F.J., The Select Committee on Energy Independence and Global Warming, Letters from the Committee, Sensenbrenner Urges IPCC to Exclude Climategate Scientists?, December 7, 2009, http://republicans.globalwarming.sensenbrenner.house.gov/press/PRArticle.aspx?NewsID=2749

Shabecoff, P., Bush Asks Cautious Response to Threat of Global Warming, *New York Times*, February 6, 1990; http://www.nytimes.com/1990/02/06/us/bush-asks-cautious-response-to-threat-of-global-warming.html?scp=81&sq=philip+shabecoff&st=nyt

Siebenhüner, B., The Changing Role of Nation States in International Environmental Assessment – the Case of the IPCC, *Global Environmental Change* 13, 113, 2003

Singh, G., and Morales, A., Melting Himalayan Glaciers Scientist Says He Was Misquoted, *Bloomberg*, January 20, 2010; http://www.bloomberg.com/apps/news?pid=newsarchive&sid=ayg2jcmPEcVo

Smith, N., and Leiserowitz, A., The Rise of Global Warming Skepticism: Exploring Affective Image Associations in the United States over Time, *Risk Analysis* (online), April 4, 2012; http://onlinelibrary.wiley.com/doi/10.1111/j.1539-6924.2012.01801.x/full

Southeastern Legal Foundation, http://www.southeasternlegal.org/about-slf/

Stacey, S., Senator Inhofe and the Heartland Institute Roll Out Underwhelming Campaign to Slash the EPA, *Climate Progress*, November 27, 2012; http://thinkprogress.org/climate/2012/11/27/1241161/senator-inhofe-and-the-heartland-institute-roll-out-underwhelming-campaign-to-slash-the-epa/

Stocker, T.F., IPCC: Cherish It, Tweak It, or Scrap It, *Nature*, 463, 731, February 11, 2010

Thacker, P.D., The Many Travails of Ben Santer, *Environmental Science & Technology*, 5834, October 1, 2008; http://pubs.acs.org/doi/pdf/10.1021/es063000t

The Daily Climate, Climate Scientist Sues for Defamation, October 23, 2012; http://wwwp.dailyclimate.org/tdc-newsroom/2012/10/climate-scientist-sues-for-defamation

The Telegraph, Climategate: UN Panel on Climate Change to Investigate Claims, December 4, 2009; http://www.telegraph.co.uk/earth/earthnews/6724249/Climategate-UN-panel-on-climate-change-to-investigate-claims.html

UK House of Commons, Science and Technology Committee, *The Disclosure of Climate Data from the Climatic Research Unit at the University of East Anglia*, March 24, 2010; http://www.parliament.uk/business/committees/committees-archive/science-technology/s-t-cru-inquiry/

UK House of Commons, Science and Technology Committee, *The Reviews into the Climatic Research Unit E-mails at the University of East Anglia*, September 8, 2010; http://www.publications.parliament.uk/pa/cm201011/cmselect/cmsctech/444/10090802.htm

UK House of Commons, Science and Technology Committee, *The Reviews into the Climatic Research Unit E-mails at the University of East Anglia*, October 27, 2010; http://www.publications.parliament.uk/pa/cm201011/cmselect/cmsctech/uc444-ii/uc44401.htm; accessed November 2011

UK House of Commons, Science and Technology Committee, *The Reviews into the Climatic Research Unit E-mails at the University of East Anglia*, January 17, 2011; http://www.publications.parliament.uk/pa/cm201011/cmselect/cmsctech/444/44402.htm

UK House of Lords, Select Committee on Economic Affairs, *The Economics of Climate Change*, July 6, 2005; http://www.publications.parliament.uk/pa/ld200506/ldselect/ldeconaf/12/12i.pdf

UK Parliament, *GOVERNMENT RESPONSE TO THE SECOND REPORT OF THE LORDS ECONOMIC AFFAIRS SELECT COMMITTEE, SESSION 2005–06*, November 2005; http://www.publications.parliament.uk/pa/ld200506/ldselect/ldeconaf/71/7104.htm

UNEP, *Expert's Profiles*, Murari Lal, http://www.unep.org/experts/default.asp?page=profiles&l=en&expertID=235

Union of Concerned Scientists, *Scientific Integrity, Timeline: Legal Harassment of Climate Scientist Michael Mann*; http://www.ucsusa.org/scientific_integrity/abuses_of_science/va-ag-timeline.html

University of East Anglia, *Cherry-picked Phrases Explained*, November 23, 2011; http://www.uea.ac.uk/mac/comm/media/press/CRUstatements/rebuttalsandcorrections/phrasesexplained

University of East Anglia, *Letter to Steven McIntyre*, March 30, 2011; http://www.climateaudit.info/correspondence/foi/20110328_%20cru_response.pdf; accessed in November 2011; also see http://www.climateaudit.info/correspondence/foi/cru/yamal/20120427%20List%20release%20letter_047_120427%20%282%29.pdf

University of East Anglia, *Release of Climate emails – November 2011*, November 22, 2011; http://www.uea.ac.uk/mac/comm/media/press/CRUstatements/statements/CRUnov11

US Department of Commerce, Office of the Inspector General, *Response to Sen. James Inhofe's Request to OIG to Examine Issues Related to Internet Posting of Email Exchanges Taken from the Climatic Research Unit of the University of East Anglia, UK*, February 18, 2011; http://www.oig.doc.gov/Pages/Response-to-Sen.-James-Inhofe%27s-Request-to-OIG-to-Examine-Issues-Related-to-Internet-Posting-of-Email-Exchanges.aspx

U.S. House of Representatives, Full Committee Hearing – Climate Change: Examining the Process Used to Create Science and Policy, March 31, 2011; http://science.house.gov/search/node/richard%20muller

US House of Representatives, Hearing Before the Select Committee on Energy Independence and Global Warming, 111th Congress, First Session, Serial No. 111–13, December 2, 2009; http://globalwarming.house.gov/pubs?id=0014; accessed February 2012

US House of Representatives, Hearing Before the Subcommittee of Energy and Environment, Committee on Science, 104th Congress, First Session, Serial No. 35, November 16, 1995; http://ia700302.us.archive.org/27/items/scientificintegr111695unit/scientificintegr111695unit.pdf

US House of Representatives, Hearing Before the Subcommittee of Energy and Environment, Committee of Science and Technology, 111th Congress, Second Session, Serial No. 111–14, November 17, 2010; http://www.gpo.gov/fdsys/pkg/CHRG-111hhrg62618/html/CHRG-111hhrg62618.htm

US Senate Committee on Environment & Public Works Hearing Statements, *The Role of Science in Environmental Policy-Making*, September 28, 2005; http://epw.senate.gov/hearing_statements.cfm?id=246814

US Senate Committee on Environment & Public Works Letter, *Senate Republicans Request UN Investigation into Climategate*, December 10, 2009; http://epw.senate.gov/public/index.cfm?FuseAction=PressRoom.PressReleases&ContentRecord_id=7A95C7C2-802A-23AD-4754-D43002001493

US Senate Committee on Environment & Public Works Report, *'Consensus' Exposed: The CRU Controversy*, February 2010; http://epw.senate.gov/public/index.cfm?FuseAction=Minority.PressReleases&ContentRecord_id=fb6d4083-802a-23ad-46e8-c5c098e22aa1&Region_id=&Issue_id=

US Senate Minority Report, *More than 700 International Scientists Dissent Over Man-Made Global Warming Claims. Scientists Continue to Debunk "Consensus" in 2008 & 2009*, original release December 22, 2008; updated March 16, 2009; http://epw.senate.gov/public/index.cfm?FuseAction=Files.View&FileStore_id=83947f5d-d84a-4a84-ad5d-6e2d71db52d9

Weber, E.U., and Stern, P.C., Public Understanding of Climate Change in the United States, *American Psychologist*, 66 (4), 315, 2011

Webster, B., and Pagnamenta, R., U.N. Must Investigate Warming 'Bias', says Former Climate Chief, *The Sunday Times*, February 15, 2010

Winter, M., Feds Clear Penn State Researcher in 'Climategate' Flap, *USA Today*, August 23, 2011; http://content.usatoday.com/communities/ondeadline/post/2011/08/feds-clear-penn-state-researcher-in-climategate-dispute/1

WWF, *An Overview of Glaciers, Glacier Retreat, and Subsequent Impacts in Nepal, India and China*, March 1, 2005; http://www.wwf.org.uk/wwf_articles.cfm?unewsid=2406

Zorita, E., IPCC: Cherish It, Tweak It, or Scrap It, *Nature*, 463, 731, February 11, 2010

Notes

1 IPCC, *Second Assessment Report, Working Group I: Scientific Assessment of Climate Change*, pp. xii, 4.
2 Ibid., p. 5
3 Edwards and Schneider, 1997, pp. 9–10; Schneider, 2009, pp. 138–42.
4 Schneider, 2009, p. 138.
5 Ibid., p. 140.
6 IPCC, *Report of the Eleventh Session of the Intergovernmental Panel on Climate Change*, p. 2.
7 Masood, June 13, 1996.
8 Seitz, June 12, 1996.
9 Santer *et al.*, June 25, 1996.
10 Bolin, 2007, pp. 127–33, and references therein.
11 For a review, see Lahsen, 1999.
12 Edwards and Schneider, 1997, p. 5.
13 Masood, June 13, 1996; also see Tracker 2008.
14 Anon, *Nature*, June 13, 1996; Bolin expressed concern with some of the editorials appearing in this journal (Bolin, 2007, pp. 107, 131).
15 Edwards and Schneider, 1997, pp. 6–7.
16 Siebenhüner, 2003, p. 417.
17 IPCC, *Report of the Thirteenth Session of the Intergovernmental Panel on Climate Change*, Appendix B, p. 3.
18 IPCC, *Report of the Nineteenth Session of the Intergovernmental Panel on Climate Change, Evaluation of the Work of Working Group III and Its Technical Support Unit from 1998–2002: Lessons for the Future*, pp. 6–7.
19 IPCC, *Report of the Seventeenth Session of the Intergovernmental Panel on Climate Change, Future Work Programme of the IPCC*, Addendum 1, pp. 1–2.
20 IPCC, *Report of the Nineteenth Session of the Intergovernmental Panel on Climate Change, Evaluation of the Work of Working Group III and Its Technical Support Unit from 1998–2002: Lessons for the Future*, p. 2.
21 IPCC, *Report of the 31st Session of the IPCC*.
22 *Earth Negotiations Bulletin*, 12 (441), p. 12, November 1, 2009.
23 Nobelprize.org; http://nobelprize.org/nobel_prizes/peace/laureates/2007/press.html
24 IPCC, *Report of the 29th Session of the IPCC*, pp. 11–13; IPCC, *Report of the 29th Session of the IPCC, Address by Ban Ki-moon*.
25 IPCC, *Report of the 29th Session of the IPCC*, p. 22.
26 Ibid., p. 11; IPCC, *Report of the 29th Session of the IPCC, Address by Ban Ki-moon*.
27 IPCC, *Review by Dutch Government*, Press Release.
28 IPCC, *Fourth Assessment Report, Working Group II, Impacts, Adaptations and Vulnerability*, 2007, p. 547.
29 Netherlands Environmental Assessment Agency, 2010, pp. 11, 24.
30 See http://www.pbl.nl/en/
31 Netherlands Environmental Assessment Agency, 2010, Annex B, pp. 86–8.
32 Ibid., p. 5.
33 Ibid., p. 6.
34 Ibid., pp. 12, 25–6.
35 For the detailed analysis see ibid., Part II, pp. 51–82; see Annex A, pp. 83–5, for the regional impacts.

36 Ibid., pp. 9, 33–5.
37 Ibid., pp. 35–7.
38 Ibid., p. 10; also see pp. 35, 55.
39 Ibid, pp. 9, 41, 43.
40 Ibid., pp. 45–50.
41 Ibid., p. 100; also see p. 9.
42 Ibid., Annex C, pp. 89–90.
43 Netherlands Environmental Assessment Agency, 2010, p. 24, Annex C; also see *EPA, Denial of Petitions for Reconsideration of the Endangerment and Cause or Contribute Findings for Greenhouse Gases under Section 202(a) of the Clean Air Act*, last updated September 9, 2013, see Response to Petitions, Vol. 2, p. 8.
44 Netherlands Environmental Assessment Agency, 2010, p. 15.
45 Ibid., p. 90.
46 *Earth Negotiations Bulletin*, 12 (384), p. 3, September 6, 2008.
47 InterAcademy Council, 2010, p. 22; IPCC, *Draft Report of the 32nd Session of the IPCC*, Annex 3, pp. 61, 64.
48 IPCC, *Fourth Assessment Report, Working Group II, Impacts, Adaptations and Vulnerability*, 2007, p. 493; it is also mentioned in the Technical Summary, p. 59, and in a summary map on p. 481.
49 IPCC, *Fourth Assessment Report of Working Group II, Impacts, Adaptations and Vulnerability, Executive Summary*, p. 59 (see Box TS.6); also see p. 49.
50 IPCC, *Fourth Assessment Report, Working Group II: Impacts, Adaptations and Vulnerability, Government and Expert Review of Second Order Draft, Specific Comments, Government Review Comments, Summary for Policymakers*, G-SPM-643, p. 92.
51 Bagla, November 13, 2009, p. 925.
52 Raina, November 12, 2009.
53 For his reporting of the Himalayan glacier issue, Bagla was awarded the David Perlman Award for Excellence in Science Journalism by the American Geophysical Union.
54 Bagla, November 9, 2009.
55 Ramesh, November 9, 2009.
56 *Financial Post*, November 30, 2012.
57 Bagla, November 13, 2009, p. 924; on December 4, *Science* also reported on the climategate scandal.
58 Brainard, September 14, 2010.
59 Bagla, December 5, 2009.
60 IPCC, *IPCC Statement on the Melting of Himalayan Glaciers*, 2010, Statements.
61 Bagla, January 29, 2010, p. 511.
62 Ibid., p. 510.
63 *Hindustantimes*, February 4, 2010.
64 Zorita, February 11, 2010, p. 731.
65 Stocker, February 11, 2010, p. 731.
66 Price, February 11, 2010, p. 732.
67 Christy, February 11, 2010, p. 732; another recommendation from an IPCC member, N.H. Ravindranath, was published in July. His suggestions included focusing more on regional change, shortening the Assessment Report cycle time, and improving the process for correcting errors, Ravindranath, 2010.
68 Hulme, February 11, 2010, p. 730.
69 Hulme, February 5, 2010.
70 Webster and Pagnamenta, February 15, 2010.
71 Revkin, August 3, 2009.
72 Christy, February 11, 2010, p. 732; InterAcademy Council, 2010, pp. 68–9.
73 Federal Register, August 13, 2010, p. 49578; also see p. 49557; the CRU issue is mentioned throughout the Federal Register; see, for example, p. 49580.
74 Note that the EPA often refers to the wrong name for the CRU as it is the Climatic Research Unit.

75 EPA, *EPA's Response to Reconsider the Endangerment and Cause or Contribute Findings for Greenhouse Gases under Section 202(a) of the Clean Air Act*, August 13, 2010, Vol. 2, p. 6.
76 Southeastern Legal Foundation, http://www.southeasternlegal.org/about-slf/
77 *EPA, Denial of Petitions for Reconsideration of the Endangerment and Cause or Contribute Findings for Greenhouse Gases under Section 202(a) of the Clean Air Act*, last updated September 9, 2013, see *Response to Petitions*, Vol. 2, p. 11.
78 Ibid., p. 17.
79 Ibid., p. 18.
80 Ibid., p. 30.
81 Ibid., p. 60.
82 Ibid., p. 20.
83 Ibid., p. 33.
84 InterAcademy Council, 2010, p. ii.
85 IPCC, *Statement of the IPCC Chairman on the Establishment of an Independent Committee to Review IPCC Procedures*, Geneva, Statements, February 27, 2010.
86 Lean, February 24, 2010.
87 IPCC, *Statement of the IPCC Chairman on the Establishment of an Independent Committee to Review IPCC Procedures*, Geneva, Statements, February 27, 2010.
88 Lean, February 26, 2010.
89 The IAC is a part of the InterAcademy Panel on International Issues, a network of over 100 national scientific academies; Pachauri had been a member of the Study Panel for the previous IAC report, *Lighting the* Way, see: InterAcademy Council, 2007.
90 InterAcademy Council, 2010, p. iii.
91 IPCC, *Report of the 29th Session of the IPCC*, p. 11; IPCC, *Report of the 29th Session of the IPCC, Address by Ban Ki-moon*.
92 IPCC, *Launch of Independent Review of the IPCC Processes and Procedures*, Press Releases, March 10, 2010.
93 InterAcademy Council, 2010, p. vii.
94 Chapter 1 was the *Introduction* and Chapter 5 was *Conclusions*.
95 InterAcademy Council, 2010, pp. xii–xvi.
96 During the preparation of the Fourth Assessment Report (AR4), Pachauri, the Chairman, had already formed the "E-team" (Executive Team; IPCC, *Report of the 30th Session of the IPCC*, April 2009, p. 9; InterAcademy Council, 2010, p. 45) of mainly Bureau members – IPCC Chair, Vice Chairs, Working Group Co-Chairs, Secretary and the heads of the Technical Support Units. A related group arose after the InterAcademy Council (IAC) inquiry into the operations of the IPCC group called the Executive Committee (InterAcademy Council, 2010, p. 45). The Executive Committee was to act on behalf of the IPCC on urgent matters in between the Plenary Sessions, and was again composed of the senior members of the Bureau (IPCC, *Report of the 33rd Session of the IPCC, Decisions Taken with Respect to the Review of IPCC Processes and Procedures, Governance and Management*, May 10–13, 2011, p. 2). These new groups had just provided more power to the senior Bureau members. However, the IAC had wanted this group to be more open. They had recommended the inclusion of three independent members on the Executive Committee (InterAcademy Council, 2010, p. xiii). The IPCC was not willing to let outsiders into their private affairs, and the independent member proposal was rejected. Essentially the new group just extended the power of the E-team that Pachauri had already set up.
97 IPCC, *Draft Report of the 32nd Session of the IPCC*, p. 1.
98 IPCC, *Draft Report of the 32nd Session of the IPCC*, p. 2.
99 IPCC, *Fourth Assessment Report, Working Group II, Impacts, Adaptations and Vulnerability*, 2007, p. 49.
100 Ibid., p. 49.
101 WWF, March 1, 2005, p. 29.

102 Ibid., p. 1.
103 Pearce, June 5, 1999.
104 http://www.cryosphericsciences.org/about.html
105 Hasnain, 1999.
106 Bagla, November 13, 2009, p. 924.
107 Alleyne, January 17, 2010; Leake and Hastings, January 17, 2010.
108 Singh and Morales, January 20, 2010.
109 Piltz, January 19, 2010; also see *Down to Earth*, April 30, 1999, and Banerjee and Collins, February 4, 2010.
110 Piltz, January 19, 2010.
111 Information from the review process is confidential and the following information discussed in this section is provided by the IPCC with the following disclaimer: "Drafts of IPCC Reports and Technical Papers, review comments, and the author responses to those comments are made available after the acceptance of the Reports and Technical Papers by the IPCC Plenary and the finalisation of the respective publication. Drafts, review comments, and author responses are confidential and may not be cited, quoted or distributed before publication of the final report (see: http://ipcc.ch/pdf/ipcc-principles/ipcc-principles-appendix-a.pdf, Section 4.1 and clarification in decision 8 on procedures taken at the 33rd Session of the Panel (see: http://www.ipcc.ch/meetings/session33/ipcc_p33_decisions_taken_procedures.pdf). These are pre-decisional materials and may not be cited or quoted as scientific literature. Only the approved, adopted and accepted reports may be cited or quoted. During the multi-stage review process, expert reviewers and governments are invited to comment on the accuracy and completeness of the scientific/technical/socio-economic content and the overall balance of the drafts. Therefore, review comments and author responses should be considered within the context of the final report. They cannot be represented as results of the IPCC assessment, and are not to be altered."
112 IPCC, *Fourth Assessment Report, Working Group II: Impacts, Adaptations and Vulnerability, Government and Expert Review of Second Order Draft, Specific Comments, Expert Review Comments, Chapter 10*, E-10-472, p. 51.
113 Ibid.
114 InterAcademy Council, 2010, p. 15.
115 Rose, January 24, 2010; also see Banerjee and Collins, February 4, 2010.
116 IPCC, *Fourth Assessment Report, Working Group II: Impacts, Adaptations and Vulnerability, Government and Expert Review of Second Order Draft, Specific Comments, Expert Review Comments, Executive Summary*, August 2006, E-TS-318, p. 46.
117 IPCC, *Fourth Assessment Report, Working Group II, Impacts, Adaptations and Vulnerability*, 2007, 2007, p. 360.
118 Leemans, 2008, pp. 14, 16; Siebenhüner, 2003, p. 417.
119 IPCC, *Fourth Assessment Report, Working Group II: Impacts, Adaptations and Vulnerability, Government and Expert Review of Second Order Draft, Specific Comments, Expert Review Comments, Chapter 10*, August 2006, E-10-468, p. 50.
120 InterAcademy Council, 2010, p. 22.
121 Ibid., E-10-466, p. 50.
122 InterAcademy Council, 2010, p. 22.
123 IPCC, *Fourth Assessment Report, Working Group II: Impacts, Adaptations and Vulnerability, Government and Expert Review of Second Order Draft, Specific Comments, Government Review Comments, Chapter 10*, August 2006, G-10-120, pp. 14–15.
124 IPCC, *Fourth Assessment Report, Working Group II: Impacts, Adaptations and Vulnerability, Government and Expert Review of Second Order Draft, Specific Comments, Government Review Comments, Summary for Policymakers*, August 2006, G-SPM-644, p. 92.

125 Ibid., G-SPM-643, p. 92.
126 InterAcademy Council, 2010, p. 22.
127 Bagla, January 29, 2010, p. 510.
128 Pachauri, May 14, 2010; during the "Discussion" period that followed his presentation.
129 InterAcademy Council, 2010, pp. xiv, 18, 22, 60–1.
130 Rose, January 24, 2010.
131 UNEP, *Expert's Profiles, Murari Lal*.
132 Bagla, November 13, 2009, p. 925.
133 Rose, January 24, 2010; also see Banerjee and Collins, February 4, 2010.
134 Rose, January 24, 2010.
135 Ibid.
136 Pearce, January 11, 2010.
137 Adam, February 14, 2010.
138 Rose, January 24, 2010.
139 InterAcademy Council, 2010, p. 48.
140 Booker, January 23, 2010.
141 McCarthy, December 7, 2011.
142 IPCC, *Fourth Assessment Report, Working Group II, Impacts, Adaptations and Vulnerability, Errata*, 2007, p. 494.
143 See IPCC, *Draft Report of the 32nd Session of the IPCC*, October 11–14, 2010, p. 11; Annex 3, p. 61.
144 Pachauri, May 14, 2010.
145 Patwardhan, January 7, 2010, slide 9.
146 InterAcademy Council, 2010, p. 20.
147 Netherlands Environmental Assessment Agency, 2010, p. 25.
148 InterAcademy Council, 2010, p. 70.
149 InterAcademy Council, 2010, pp. 20–1.
150 IPCC, *Report of the 28th Session of the IPCC Bureau*, p. 2.
151 InterAcademy Council, 2010, p. xvi.
152 IPCC, *Report of the 34th Session of the IPCC, Review of IPCC Processes and Procedures, Procedures: Adoption of the revised "Appendix A to the Principles Governing IPCC Work: Procedures for the Preparation, Review, Acceptance, Adoption, Approval and Publications of IPCC Reports" Role of Review Editors – Revised Guidance Notes*, November 18–19, 2011, Add. 1, p. 2.
153 IPCC, *Draft Report of the 32nd Session of the IPCC*, October 11–14, 2010, p. 14.
154 IPCC, *Report of the 33rd Session of the IPCC, Review of the IPCC Processes and Procedures, Proposal by the Task Group on Procedures*, May 10–13, 2011, p. 5.
155 IPCC, *Report of the 33rd Session of the IPCC, Decisions Taken with Respect to the Review of IPCC Processes and Procedures, Procedures*, May 10–13, 2011, p. 2.
156 InterAcademy Council, 2010, p. xvi.
157 Ibid., p. 63.
158 *Earth Negotiations Bulletin*, 12 (486), p. 9, October 18, 2010.
159 IPCC, *Draft Report of the 32nd Session of the IPCC*, October 11–14, 2010, p. 9; Annex 3, p. 66; p. 16; Annex 3; IPCC, *Decisions Taken by the Panel at the 32nd Session*.
160 IPCC, *Report of the 33rd Session of the IPCC, Review of the IPCC Processes and Procedures, Draft Proposal by the Task Group on Governance and Management, Terms of Reference for the Bureau and Bureau Members*, May 10–13, 2011, p. 2.
161 IPCC, *Report of the 33rd Session of the IPCC, Decisions Taken with Respect to the Review of IPCC Processes and Procedures, Governance and Management*, May 10–13, 2011.
162 Ibid., p. 6; also see IPCC, *Report of the 34th Session of the IPCC, Further Work to Adopting Revisions to "Appendix C of the Principles Governing IPCC Work: Rules of Procedures for the Election of the IPCC Bureau and Any Task Force Bureau,"* November 18–19, 2011, p. 13.

163 *Earth Negotiations Bulletin*, 12 (547), June 12, 2012, p. 10.
164 How the records ended on the internet have not been confirmed, see Revkin, March 1, 2011; also see Arthur, February 5, 2010.
165 Russell *et al.*, July 2010, p. 26.
166 University of East Anglia, *Cherry-picked Phrases Explained*, November 23, 2011.
167 University of East Anglia, *Release of emails – November 22, 2011*.
168 Russell *et al.*, July 2010, p. 26.
169 McCrae, December 2, 2009; Girling, February 7, 2010; Adam, November 15, 2010.
170 Ravilious, December 8, 2009.
171 UK House of Commons, March 24, 2010, p. 14.
172 Russell *et al.*, July 2010, p. 37; UK House of Commons, March 24, 2010, p. 12.
173 See Russell *et al.*, July 2010, pp. 29–34, 71, 91; Bolin, 2007, p. 167.
174 Russell *et al.*, July 2010, pp. 57–9, 62.
175 Mann *et al.*, 1999.
176 For a general discussion of the debate see Monastersky, 2006.
177 BBC News, November 20, 2009.
178 Hickman and Randeson, November 20, 2009.
179 Delingpole, November 20, 2009.
180 See, for example, Revkin, November 21, 2009.
181 Kintisch, December 4, 2009.
182 UK House of Commons, March 24, 2010, p. 42.
183 Morano, November 29, 2009.
184 Revkin, November 27, 2009; also see Hulme and Ravetz, December 1, 2009.
185 *The Telegraph*, December 4, 2009; also see IPCC, *IPCC Statement on News Reports Regarding Hacking of the East Anglia University Email Communications*, December 4, 2009.
186 Absai, December 13, 2009.
187 Bagla, January 29, 2010, p. 510.
188 BBC News, January 9, 2004.
189 May, January 27, 2005.
190 UK House of Lords, July 6, 2005, p. 6.
191 Ibid., p. 16.
192 Ibid., pp. 18–20.
193 Ibid., p. 23.
194 Ibid., p. 55.
195 Ibid., p. 73.
196 UK Parliament, November 2005.
197 UK House of Commons, March 24, 2010, p. 8.
198 UK House of Commons, January 17, 2011.
199 Ibid.
200 Russell *et al.*, July 2010, pp. 55–6, 63, 70–85; UK House of Commons, March 24, 2010, pp. 21–2.
201 UK House of Commons, January 17, 2011.
202 UK House of Commons, October 27, 2010, pp. 36–41, 49.
203 Russell *et al.*, July 2010, p. 77.
204 IPCC, *IPCC Statement on News Reports Regarding Hacking of the East Anglia University Email Communications*, December 4, 2009.
205 Russell *et al.*, July 2010, pp. 11–14.
206 UK House of Commons, January 17, 2011.
207 UK House of Commons, March 24, 2010, p. 45.
208 Oxburgh *et al.*, April 19, 2010, pp. 2, 3, 5; this criticism was also brought up during the EPA review, EPA, *EPA's Response to Reconsider the Endangerment and Cause or Contribute Findings for Greenhouse Gases under Section 202(a) of the Clean Air Act*, August 13, 2010, Vol. 2, p. 47, Comment 2–66.
209 Oxburgh *et al.*, April 19, 2010, pp. 1, 3, 5.

210 Ibid., p. 5.
211 UK House of Commons, September 8, 2010.
212 UK House of Commons, January 17, 2011.
213 UK House of Commons, October 27, 2010.
214 UK House of Commons, January 17, 2011.
215 UK House of Commons, *Key Findings*, January 17, 2011.
216 UK House of Commons, March 24, 2010, p. 37.
217 Montford, September 2010, p. 7.
218 Ibid., p. 54; also see McKitrick, September 2010.
219 UK House of Commons, March 24, 2010, pp. 14, 16, 32–5, 42–3; Russell *et al.*, July 2010, pp. 94–5.
220 UK House of Commons, March 24, 2010, pp. 44, 50.
221 UK House of Commons, Science and Technology Committee, *Moving Forward at UEA*, January 17, 2011.
222 UK House of Commons, March 24, 2010, pp. 14–16; for an earlier example, see Bolin, 2007, p. 134.
223 Russell *et al.*, July 2010, p. 35; also see UK House of Commons, March 24, 2010, p. 47.
224 UK House of Commons, *The Disclosure of Climate Data from the Climatic Research Unit at the University of East Anglia*, March 24, 2010, p. 46.
225 University of East Anglia, March 30, 2011.
226 Jha, May 25, 2011; UK House of Commons, September 6, 2010; also see Schneider, 2009, pp. 146–8.
227 Russell *et al.*, July 2010, pp. 90, 95.
228 Foley *et al.*, February 3, 2010, p. 1.
229 Assmann *et al.*, June 4, 2010, p. 3, Foley *et al.*, February 3, 2010, p. 3.
230 Assmann *et al.*, June 4, 2010, p. 5, Foley *et al.*, February 3, 2010, pp. 5–9.
231 Assmann *et al.*, June 4, 2010, p. 19; for the media response to the investigation see Dawson, March 18, 2010.
232 Winter, August 23, 2011.
233 National Science Foundation, August 15, 2011.
234 Ibid.
235 Cuccinelli, February 17, 2010.
236 Russell, April 27, 2010.
237 Heiderman, May 9, 2010.
238 Anon, Science Subpoenaed, May 13, 2010.
239 *The Daily Climate*, October 23, 2012.
240 Union of Concerned Scientists.
241 *The Daily Climate*, October 23, 2012.
242 See for example McCright and Dunlap, 2003, p. 361.
243 US House of Representatives, November 16, 1995.
244 Lahsen, 1999, p. 133.
245 National Research Council, National Academy of Sciences, 2001, p. 4.
246 Maassarani, 2007, p. 62.
247 Monastersky, 2006.
248 US House of Representatives, 2009, p. 5.
249 Ibid., p. 6.
250 Ibid., p. 58.
251 Ibid., pp. 13–14.
252 Ibid., pp. 19–20.
253 Ibid., p. 54.
254 Ibid., pp. 52, 68.
255 Ibid., p. 53.
256 Sensenbrenner, December 7, 2009.
257 Bravender, September 23, 2010.

258 US House of Representatives, November 17, 2010.
259 US House of Representatives, March 21, 2011.
260 Lacey, November 28, 2012.
261 Ibid.
262 Inhofe, *Science of Climate Change*, July 28, 2003, 19933.
263 Ibid., 19937.
264 Ibid., 19938.
265 Ibid., 19941; also see Revkin, August 5, 2003.
266 Inhofe, *Science of Climate Change*, July 28, 2003, 19943.
267 Inhofe, October 11, 2004, p. S11292.
268 Inhofe, *The Facts and Science of Climate Change*, n.d., p. 9; Inhofe, July 29, 2003.
269 Inhofe, *The Facts and Science of Climate Change*, n.d., p. 17.
270 Ibid., p. 20.
271 US Senate Committee on Environment & Public Works Hearing Statements, *The Role of Science in Environmental Policy-Making, Statement of Michael Crighton*, September 28, 2005.
272 Nisbet, 2009, p. 19.
273 US Senate Minority Report, March 16, 2009.
274 McCullagh, November 24, 2009.
275 US Senate Committee on Environment & Public Works Letter, December 10, 2009.
276 Barrasso, February 4, 2010.
277 US Senate Committee on Environment & Public Works Report, February 2010, p. 6.
278 Ibid., p. 24.
279 Ibid., p. 29.
280 Ibid., p. 35; the critical report of the skeptic Montford (Montford, September 2010, p. 10) contains only seven names (six of whom are in the Group of Seventeen). Michael Mann holds the dubious distinction of being one of the most politically harassed scientists in American history.
281 Goldenberg, October 31, 2010.
282 US Department of Commerce, Office of the Inspector General, *Response to Sen. James Inhofe' s Request to OIG to Examine Issues Related to Internet Posting of Email Exchanges Taken from the Climatic Research Unit of the University of East Anglia, UK, Enclosure*, February 18, 2011, p. 3.
283 Ibid.
284 *Response to Sen. James Inhofe's Request to OIG to Examine Issues Related to Internet Posting of Email Exchanges Taken from the Climatic Research Unit of the University of East Anglia, UK, Summary of Results*, February 18, 2011, pp. 3, 13–17.
285 Pielke, November 22, 2009.
286 US Department of Commerce, Office of the Inspector General, *Response to Sen. James Inhofe' s Request to OIG to Examine Issues Related to Internet Posting of Email Exchanges Taken from the Climatic Research Unit of the University of East Anglia, UK, Enclosure*, February 18, 2011, p. 8.
287 Ibid.
288 Ibid., p. 17.
289 Stacey, November 27, 2012.
290 Jarvis, June 19, 2013; other Republicans made this list as well.
291 McCright and Dunlap, 2010, p. 119.
292 Kintisch, December 4, 2009, p. 1329; also see Russell *et al.*, July 2010, p. 31; Weber and Stern, 2011.
293 Anon., Climate of Fear, March 11, 2010.
294 Ibid.
295 Gleick *et al.*, 2010, pp. 689–90.
296 Painter, 2011, p. 1; also see Painter and Ashe, 2012.

297 Nisbet, 2009, p. 14; Smith and Leiserowitz, April 4, 2012; Leiserowitz *et al.*, May 2011, p. 56; Weber and Stern, 2011, p. 322.
298 McCright and Dunlap, 2011. pp. 179–80.
299 Brulle *et al.*, February 2, 2012; also see *Science Daily*, October 4, 2012.
300 IPCC, Opening Statement by Dr Rajendra Pachauri, Chairman of the Intergovernmental Panel on Climate Change at a Press Conference at the United Nations in New York, Statements, 2010.
301 Schneider, 1991, p. 25.
302 Schneider, 2009, p. 109.
303 Ibid., p. 120.
304 Bolin, 2007, p. 73.
305 Ibid., p. 63.
306 McCright and Dunlap, 2010, p. 115; a critique of the climate change skeptics is presented by Oreskes and Conway, June 2010.
307 See Leggett, April 25, 2006.
308 The opposition of the GCC knew few bounds, including colluding with OPEC states to undermine the messages of the IPCC; see Shabecoff, February 6, 1990; Edwards and Schneider, 1997, p. 10; Schneider, 2009, pp. 122–3; Masood, June 13, 1996; Bolin, 2007, pp. 83, 103, 111, 123; Gelbspan, 1997.
309 Bolin, 2007, p. 75.
310 Ibid., pp. 72–3.
311 Nisbet, 2009, pp. 18–19.
312 Bolin, B., Science and Policymaking, 1994, p. 27; also see Bolin, Next Step, 1994.
313 Russell *et al.*, July 2010, p. 15; also see pp. 30, 41–2.
314 Antilla, 2005, p. 339; Weber and Stern, 2011, p. 320.
315 All web-based references were confirmed (accessed) in June 2013, unless otherwise stated.
316 For the report of the 32nd Session and documents for the 33rd Sessions see: http://www.ipcc.ch/scripts/_session_template.php?page=_33ipcc.htm
317 This and similar references below can be found at: http://ipcc-wg2.gov/publications/AR4/ar4review.html
318 IPCC Statements can be found at: http://www.ipcc.ch/news_and_events/press_information.shtml
319 IPCC Press Releases can be found at: http://www.ipcc.ch/news_and_events/press_information.shtml
320 IPCC Plenary Session minutes and supporting documents can be found at: http://www.ipcc.ch/meeting_documentation/meeting_documentation_ipcc_sessions_and_ipcc_wgs_sessions.shtml

3   The players

Sitting on the bench – US climate change policy

The US has generally been a reluctant member of the international community in the climate change discussions, as carbon targets were viewed as being detrimental to their economy.[1] And even when they did participate, the Americans were viewed as a disruptive force in the discussions.[2] The traditional blocking mechanism was put forth by the government; more research on the science and economics of climate change was needed before any legally binding decisions could be considered. The "paralysis-by-analysis" strategy on climate change by American administrations has worked for three decades (and we are now into a fourth).[3]

The US had an unexpected ally in their mission to hamper the negotiations. OPEC[4] was, for once, helping the Americans by also undermining the negotiating process. The oil-producer coalition was simply protecting its own economic interests as any policies were bound to reduce consumption of fossil fuels, including oil. Trouble started right at the beginning when there was no agreement to determine the Rules of Procedures for the negotiating process. This usually routine bureaucratic decision had been repeatedly blocked by OPEC.[5] Developed nations were also not entirely displeased with this outcome.[6] They were leery of being forced into policies by a majority voting system, as they were far outnumbered by the developing nations. Consensus essentially gave them a veto position to use whenever they wanted. This also played into the strategy of the Bush Administration as it bogged down the entire process. However, even the subsequent Clinton Administration was not "enthusiastic"[7] about altering the procedure to majority rule. Attempts to change the consensus approach were still being blocked two decades later; at Durban in December 2011, Bolivia, Venezuela and Saudi Arabia vetoed any changes.[8] There was some irony in the venue of the COP meeting at the end of the Kyoto season being Doha, Qatar.

During the Eisenhower Administration, in 1956, Roger Revelle started off the interaction between climate science and American politics when he presented his concerns on global warming to the US Congress.[9] The carbon game was afoot, as scientists attempted to draw politicians into action on an issue that only the scientists could fully appreciate. The issue of climate change and public policy was raised again in 1963 at a conference by the Conservation Foundation.[10] The first warning on climate change to the White House by an eminent group of scholars took place in 1965 by the President's Science Advisory Committee for Lyndon Johnson (1908–73); a report of the Environmental Pollution Panel, which included contributions from David Keeling and Roger Revelle, stated:[11]

> Throughout most of the half-million years of man's existence on Earth, his fuels consisted of wood and other remains of plants which had grown only a few years before they were burned. The effect of this burning on the content of the atmospheric carbon dioxide was negligible, because it only slightly speeded up the natural decay processes that continually recycle carbon from the biosphere to the atmosphere. During the last few centuries, however, man has begun to burn the fossil fuels that were locked in the sedimentary rocks over five hundred million years and this combustion is measurably increasing atmospheric carbon dioxide …
>
> Through his worldwide industrial civilization, Man is unwittingly conducting a vast geophysical experiment … By the year 2000 the increase in atmospheric CO_2 will be close to 25%. This may be sufficient to produce measureable and perhaps marked changes in climate … The climatic changes that may be produced by the increased CO_2 could be deleterious from the point of view of human beings.

On July 30, 1979, Senator Abraham A. Ribicoff (1910–98) of Connecticut, and Edmund Muskie (1914–96) of Maine, held a Senate hearing called "Carbon Dioxide Accumulation in the Atmosphere, Synthetic Fuels and Energy Policy," where Revelle and Schneider presented.[12] Another important workshop took place at Wood Hole, Massachusetts, under the auspices of the National Academy of Sciences,[13] where Keeling and Revelle were again among the co-authors of the report that was sent to President Carter (1924–):[14]

> Man is setting into motion a series of events that seem certain to cause a significant warming of world climates over the next decades unless mitigating steps are taken immediately … Enlightened policies in the management of fossil fuels and forest can delay or avoid these changes, but the time for implementing the policies is fast passing.
>
> The potential disruptions are sufficiently great to warrant the incorporation of the CO_2 problem into all considerations of policy in the development of energy.

Carter "was the first politician to promote an industrial revolution with renewables."[15] Among his actions was the formation of the Solar Energy Research Institute.

"Acting" on climate change

As a symbolic gesture, President Carter installed solar panels at the White House in 1979. Seven years later, they were removed by President Reagan (1911–2004).[16]

He complained of "solar socialism" and is claimed to have said "go away with this shit of renewables."[17]

Of historic interest is the role that the Reagan Administration played in the formation of the IPCC. In a rare insider look at US politics on climate change, the EPA officials (and both members of the IPCC) Alan D. Hecht, the Principal Deputy Assistant Administrator of the Office of International Activities,[18] and Dennis Tirpak, Director of the Global Change Division, Office of Policy and Planning,[19] reported on the Machiavellian gamesmanship within the inner circles of government:[20]

> It may not be widely appreciated that UNEP's efforts to promote a convention on climate specifically triggered the development of the Intergovernmental Panel on Climate Change (IPCC). In response to the recommendations of the Villach report, UNEP Executive Director Moustafa Tolba sent a letter to Secretary of State George Schultz urging the U.S. to take appropriate policy actions ... The majority of federal agencies present did not support negotiations of a climate convention. The mood of senior officials then in Washington was that the underlying scientific evidence for global warming was inconsistent, contradictory and did not justify policy actions that likely would be expensive. The Department of Energy felt strongly that the Villach report was inadequate because it was not prepared by government officials.
>
> EPA and the Department of State supported the idea of a convention and suggested that perhaps it was timely for governments to prepare an international scientific assessment, especially in light of conflicting scientific evidence. This debate within the Policy Board led to the U.S. proposal for 'an intergovernmental mechanism', to conduct a government-led, scientific assessment of the climate change issue. This 'mechanism', when later presented to WMO and UNEP governing bodies in the form of intergovernmental resolutions, became the Intergovernmental Panel on Climate Change (IPCC) ...
>
> At a time when it was difficult to get interagency agreement on any action, there was convergence (for different reasons) around the concept of an international scientific assessment. Some U.S. agencies clearly saw this as a means to 'buy time' before engaging in serious policy decisions ...
>
> The authors participated in this meeting and have a clear record of the discussion. We are asserting that an element of the U.S. support for the IPCC concept, at least by some agencies, was based on the view that the IPCC would slow down the push for a climate convention.

A later commentary stated:[21]

> So the Republican administration under Ronald Reagan worked the UN corridors, lobbying against Tolba's plan and pushing for the creation of an Intergovernmental Panel on Climate Change, which was duly established by the UN General Assembly in 1988.

The purpose of the IPCC was to put scientists back in the cages they had briefly escaped from at Villach, and to this day the IPCC's members remain government nominees.

In June 1988, the Americans had been reluctant participants at the Toronto Conference, where they claimed that the Canadian delegates were "very heavy handed in preparing the final report."[22] Despite the growing scientific evidence of the threat of climate change, the US did not see the need to play the carbon game. In 1988, Richard J. Smith of the State Department told US Congress that such discussions were "premature."[23] The government "had emphasized on more research before acting."[24] Tolba would later warn the IPCC that it had to remove "excuses for inaction."[25] While no nations were specifically mentioned, the message was obviously directed at one country in particular.

Back in 1983, the EPA had issued a report (*Can We Delay a Greenhouse Warming?*) which highlighted the growing conflict between coal for energy production and climate change. And four years later, President Reagan had signed the Climate Protection Act, which called upon the EPA to develop a plan to mitigate climate change.

Two Democrats who would feature prominently during the following Clinton Administration were active during the Reagan years – Representative, and later Senator, Al Gore (1948–) of Tennessee, and Senator Tim Wirth (1939–) of Colorado. In July 1981, Gore had held hearings at the House of Representatives on climate change,[26] and had been one of a prominent group of politicians that requested information on climate change from the National Academy of Sciences on June 2, 1986.[27]

On June 23, 1988, the Senate Committee on Energy and Natural Resources invited the expert climate change scientist, Jim Hansen (1941–), who warned: "Global warming has reached a level such that we can ascribe with a high degree of confidence a cause and effect relationship between the greenhouse effect and observed warming … It is already happening now."[28] The Committee Chairman, Senator Tim Wirth, concluded:

> As I read it, the scientific evidence is compelling: the global climate is changing as the Earth's atmosphere gets warmer. Now, the Congress must begin to consider how we are going to slow or halt that warming trend and how we are going to cope with the changes that may already be inevitable.

Wirth had been a promoter of the Crocker–Dales flexible mechanisms to deal with acid rain. He would go on to direct the US climate change portfolio during Clinton's presidency.

AmBush at the INC

The architect of the American doctrine on climate change was George H. W. Bush (1924–). In September 1988, during his election campaign, there had been encouraging words from the candidate:[29]

Those who think we are powerless to do anything about the 'greenhouse effect' are forgetting about the 'White House effect,' … In my first year in office, I will convene a global conference on the environment at the White House. It will include the Soviets, the Chinese … The agenda will be clear. We will talk about global warming.

And the first year of the Administration was a pleasant surprise as there was support for a framework convention on climate change. On January 30, 1989, James Baker (1930–), the Secretary of State, stated:[30]

[W]e can probably not afford to wait until all of the uncertainties have been resolved before we do act. Time will not make the problem go away … while scientists refine the state of our knowledge, we should focus immediately on prudent steps that are already justified on grounds other than climate change.

On May 12, 1989, William K. Reilly (1940–), the Administrator of the EPA, announced that the President had agreed to participate in the framework convention process.[31] And in December 1989 President Bush offered to host the first meeting of the negotiations in Washington.[32]

However, Al Gore, then the Senator from Tennessee, believed that it was too early to praise the Administration:[33]

Once again, the President has been dragged slowly and reluctantly toward the correct position when the White House should have been providing leadership. I welcome their decision finally to begin recognizing the importance of global climate change as an international issue demanding attention, but it's too early for rave reviews of a policy developed in an atmosphere of political damage control in response to Congressional pressure.

Senator Gore's suspicions were well founded, as a policy coup took place in the White House during the second year of the Bush Administration. Baker was moved out of the portfolio, officially because of his former oil ties,[34] and Reilly was forced into a "marginal" role.[35] Orchestrator of the coup was John H. Sununu (1939–), the White House Chief of Staff, who strongly opposed any treaty on climate change.[36] From the details of a story told by William Reilly of the EPA, Sununu was a climate skeptic: "[A]t the conclusion of the meeting Sununu told the Cabinet that the models on which climate change were premised were fundamentally flawed and the best atmospheric scientists had yet to become involved in climate research."[37]

The original lead US negotiator of the climate change negotiations, William A. Nitze, the Deputy Assistant Secretary of State for Environment, 1987–90, was frustrated by the influence of Sununu:[38]

This brings us to the basic explanation for the Bush administration's negative posture during the climate change negotiations – a failure of presidential leadership. President Bush's failure to come to grips with the climate issue is in large part the result of personal ignorance. In contrast to Margaret Thatcher and other world leaders, the President never took the time to be thoroughly briefed about the science, impacts or policy implications of climate change. Instead he chose to rely on the knowledge, opinions and judgment of a few key advisors, particularly his former Chief of Staff, Mr. Sununu.

Sununu and the energy industry lobby were lampooned in the 1991 movie *Naked Gun 2½: the Smell of Fear*. In the film, John Sununu (played by Peter van Norden) introduced the trade organization leaders at a White House dinner:

- Coal – Terence Baggett – Society for More Coal Energy – "SMOCE"
- Oil – Donald Fenswick – Society of Petroleum Industry Leaders – "SPIL"
- Nuclear – Arthur Dunwell – Key Atomic Benefits Office of Mankind – "KABOOM."

Under the influence of the Sununu doctrine, the policy of the Americans became one of obstructing progress of the climate change talks. At the pre-INC meetings at the Noordwijk Ministerial Conference in November 1989, and the Second World Climate Conference in Geneva a year later, the US refused to define the level of GHG emissions reductions.[39] Bush maintained this obstructionist posture despite warnings from the National Academy of Sciences, presented by the Nobel Laureate in physics, Henry Way Kendall (1926–99), in early 1990:[40]

I am saddened that the main responsibility for the state of affairs belongs to my own country. The refusal of the United States to make a commitment to reduce, or even to stabilize, its huge emissions of carbon dioxide and other greenhouse gases has been the principal roadblock to progress. In early 1990, I transmitted to President Bush an appeal for action to prevent global warming on behalf of 55 American Nobel Laureates and more than 700 members of the National Academy of Sciences.

When President Bush welcomed the IPCC to Washington, DC, on February 5, 1990, his focus was on economics:[41]

The economic considerations of our response strategies to climate change is getting intensive study here in our country, in the United States. We are developing real data on the costs of various strategies, assessing new measures, and encouraging other nations to follow suit.
   Wherever possible, we believe that market mechanisms should be applied – and that our policies must be consistent with economic growth and free market principles in all countries.

Bush did keep his promise to hold an international conference on climate change, which was also mentioned at the IPCC meeting:[42]

Therefore, this spring, the United States will host a White House conference on science and economic research on the environment – convening top officials from a representative group of nations, to bring together the three essential disciplines: science, economics, and ecology … I look forward personally to participating in this seminar and to learning from its deliberations …

Our goal continues to be matching policy commitments to emerging scientific knowledge.

On April 17, 1990, a meeting called "Science and Economics Research Related to Global Change" was held at the White House. Although Bush had promised to talk about "global warming," he failed to mention the term even once during his address to delegates at the White House conference. The Administration exerted a Machiavellian-like control over the meeting; the briefing notes for American officials directed them to avoid discussing the major issues.[43] Following a text-book strategy to derail any global agreement, the notes went on to suggest: "A better approach is to raise the many uncertainties that need to be better understood on this issue."[44]

There had been an American ambush at the conference. Though 17 nations attended, there was only one non-American speaker and most simply endorsed the US position that the science was still uncertain. No one had even invited the IPCC, until Bolin complained.[45]

US negotiators pleasantly surprised delegates at the first session of the INC in early 1991. They announced that their emissions would be below 1987 levels by 2000.[46] Furthermore, the US endorsed a plan of "meaningful action" by all nations without using a "fixed formula," and that the "financial and technological realities of all nations" should be taken into consideration. But just a month later, on March 15, 1991, the Americans had reversed their stance,[47] returning to their standard position of only voluntary and nebulous targets and rejecting firm, legal commitments.[48] The reversal was blamed on "forceful interests at work in Washington."[49]

The frustration felt by delegates to the INC is generally documented in the insider book, edited by Mintzer and Leonard.[50] The Americans had isolated themselves at the discussions,[51] and the Brazilian delegate to the INC, José Goldemberg, described the rift between the other 153 delegates and the "hard-line"[52] position of the United States. Ripert, the Chairman of the INC, feared that the Earth Summit itself was in jeopardy, and other members of the INC worried that President Bush would not even attend the conference at Rio,[53] as he had refused to confirm his attendance without a US-approved agreement.[54] In April 1992, Ripert had called delegates together in Paris to attempt to resolve the impasse. During the same month, the European Community President, Jacques Delors (1925–), came to Washington to meet with President Bush. American church leaders, leading scientists such as Stephen Schneider, and Al Gore, pressured Bush to attend the Rio Conference.[55] However, the President of the United States was blackmailing the INC negotiators to accept a watered-down resolution or else he was not going to play the game. This threat led the INC to settle for only "aims" with no legally binding targets.[56] Officials of the EPA delicately described the agreement as follows:[57]

The carefully chosen but often contorted language in the convention was the end result of more than two years of intense international negotiation and debate between the U.S. and European Community (E.C.) countries on approaches and commitments toward stabilizing greenhouse gases and developed and developing countries on overall responsibilities and commitments.

In return, President Bush agreed to attend the Rio Conference and signed the UNFCCC.[58] As Stephen Schneider aptly wrote, "Bush did the right thing for history,"[59] but there was no real deal at Rio, just vague intentions.[60] As his term was ending, in October 1992, the US became the first industrialized nation to ratify the UNFCCC.[61] The President of the United States had won the first game of the season with the American ambush.

The "Gorey" details

The year 1993 began with high hopes that the US would have a more positive approach to the climate change negotiations under the new leadership of Bill Clinton (1946–) and Al Gore, who had both been strong supporters of Rio. Joining this team on the climate change portfolio were Tim Wirth and Rafe Pomerance[62] (1946–). As Senator from Colorado, Wirth had chaired the historic Senate meeting with the climate crusader Jim Hansen of NASA in 1988, and had attended the Earth Summit in Rio as a member of the Senate observer group. He was now the Under-Secretary of State for Global Affairs, a newly created position in the Clinton–Gore Administration.

As with the previous Administration, initial actions by the Clinton Administration were encouraging. Shortly after taking office, in February 1993, President Clinton introduced the "Btu tax" on energy. Although Al Gore had favored a "carbon tax," a more general energy tax was proposed as a lesser burden on the coal industry.[63] On April 21, 1993 (Earth Day), President Clinton made the bold statement that the US would return to emissions levels of 1990 by 2000:[64]

> We must take the lead in addressing the challenge of global warming that could make our planet and its climate less hospitable and more hostile to human life. Today, I reaffirm my personal, and announce our nation's commitment to reducing our emissions of greenhouse gases to their 1990 levels by the year 2000. I am instructing my administration to produce a cost-effective plan … that can continue the trend of reduced emissions. This must be a clarion call, not for more bureaucracy or regulation or unnecessary costs, but instead for American ingenuity and creativity, to produce the best and most energy-efficient technology.

However, the proposed "Btu tax" was defeated in the Senate.[65] A weak Climate Action Plan was issued in October with only voluntary measures. The Clinton Administration retreated back to the Bush doctrine[66] of no legally binding targets or timetables.

In 1994, the Republicans gained a majority in both Houses, dashing any hope of further progress. An even greater problem was that there was a growing atmosphere of distrust between the White House and Congress, which transcended party lines. And the relationship got worse when the Administration failed to provide an economic evaluation of climate change mitigation plans after it had been repeatedly requested by Congress. An economic review was underway in 1995; the verdict was that the target of getting emissions back to 1990 levels by 2010 would take a carbon tax of $95 per ton.[67] The administration rejected the evaluation. Not until after Kyoto did Congress finally get an answer.[68]

One week before the first COP was to commence in Berlin, the US House of Representatives held its inaugural meeting on the "International Global Climate Change Negotiations" by the Subcommittee on Energy and Power, part of the powerful Committee on Commerce. The hearing, chaired by Dan Schaefer (1936–2006) of Colorado, was to present an overview of the "status of global climate change negotiations and the impact on the U.S. economy." Generally, these hearings are not for the faint of heart and, at times, resembled a political inquisition. Witnesses, often from government departments and agencies, are put on the hot seat, and, in this case, when the US negotiating team on climate change showed up at Capitol Hill, things really overheated. Tim Wirth was not available for the Subcommittee hearing as he was in the midst of preparations for the COP (April 1995). As a result, the defense of the Administration's position fell to Pomerance and Susan Tierney (1951–).

Many members of the Subcommittee were frustrated by previous dealings with the White House and were already hostile to the climate change negotiating team. Congress was obviously not pleased by the manner in which Wirth and the Administration had been handling the international negotiations. The Chairman opened the hearing by outlining three concerns:[69]

1. The American negotiating position was unclear.
2. The position of other developed nations was unknown.
3. Developing countries may not be "full participants in the treaty."

The Subcommittee was well versed on the matter, and they were aware of the pivotal issue of the phrase "common but differentiated responsibilities."[70] Pomerance presented the Administration's interpretation of the controversial phrase as: "The industrialized countries have basically to take a leadership role, but all countries have obligations to take actions under the convention."[71] The definition had been only reluctantly provided by Pomerance after being badgered by John D. Dingell (1926–), the long-standing Democratic Representative from Michigan, who would not accept waffling answers to his questions. Dingell was following his aggressive line of questioning because he feared that the US would agree to a deal in COP 1 at Berlin that would hurt US competitiveness. The answers from Pomerance did little to appease the Congressman, so Dingell brought the issue up directly with President Clinton. On March 31, 1995, the President attempted to calm the senior Representative from Michigan.[72] Nevertheless, Dingell was convinced that something was amiss and that the US would come up short in Berlin. Before COP 1, the Global Climate Coalition (GCC) lobbied heavily and presented the following warning to Congress: "Our fear is that in order to placate Germany and others, the developed nations in Berlin will let developing countries off the hook so that the forthcoming session can produce a Berlin accord."[73]

At COP 1, just as had been predicted by the GCC, the developing countries ended up with *no* emissions targets. How could the White House claim that the Berlin Mandate would not hurt the US when competing countries, such as China and India, would have no carbon targets?

Wirth was actually pleased[74] with this outcome. However, he would soon have to account for his actions. On May 19, 1995, only two months after Berlin, the Subcommittee on Energy and Power met again. Under the circumstances, it is almost comical that Wirth declared a complete victory for the US at the hearing.[75] Inexplicably, Wirth could not see any problem with the Berlin Mandate: "We did not agree, and this is nowhere in there, that other countries don't have responsibilities. They do have responsibilities. That is recognized in this language."[76] The US lead COP negotiator goes on to describe all the commitments and obligations of the developing nations. However, he was twisting the facts, for these were all administrative duties and Congress had been looking for emissions targets. Wirth then downplayed the Berlin Mandate by stating it was merely a stepping stone towards negotiations with developing countries for targets after 2000.

The members of the Subcommittee in Congress did not interpret the outcome of COP 1 in the same way as their *über*-enthusiastic lead negotiator. Dan Schaefer, Chairman of the Subcommittee, stated for the record, "This seems to make little sense."[77] Then John Dingell waded in and he was not going to let Wirth off the hook;[78] their argument soon became circular, with Wirth claiming that the US could still get developing countries on board, and Dingell stating that the Berlin Mandate precluded this from happening. But the implications of the Berlin Mandate were indisputable – the G77 and China had repeatedly stated that they would not accept emissions targets, and the Berlin Mandate could only be used to support their rigid position. John Dingell laid this out for the apparently confused Wirth:[79]

> I regard this as committing ourselves to a serious problem when it comes time to get down to discussing the next meetings, and they are going to read to you, Mr. Wirth, and they are going to say, but you have agreed that you are not going to seek to impose any new commitments upon us.

Only four months before Kyoto was to begin, on July 25, 1997, a frustrated US Senate took draconian action against the climate change agenda of the Clinton–Gore Administration. A veto, the Byrd–Hagel Resolution, was passed if the Berlin Mandate remained part of the Kyoto Protocol:[80]

> (1) the United States should not be a signatory to any protocol to, or other agreement regarding, the United Nations Framework Convention on Climate

Change of 1992, at negotiations in Kyoto in December 1997, or thereafter, which would –

(A) mandate new commitments to limit or reduce greenhouse gas emissions for the Annex I Parties, unless the protocol or other agreement also mandates new specific scheduled commitments to limit or reduce greenhouse gas emissions for Developing Country Parties within the same compliance period, or

(B) result in serious harm to the economy of the United States; and

(2) any such protocol or other agreement which would require the advice and consent of the Senate to ratification should be accompanied by a detailed explanation of any legislation or regulatory actions that may be required to implement the protocol or other agreement and should also be accompanied by an analysis of the detailed financial costs and other impacts on the economy of the United States which would be incurred by the implementation of the protocol or other agreement.

In a stunning rebuke of the White House management of climate change policy, the resolution was passed by the US Senate unanimously! At the time, Wirth officially stated that the resolution strengthened the American position at Kyoto.[81] Shortly afterwards, on November 19, just two weeks before Kyoto,[82] Wirth resigned.

His replacement, Stuart Eizenstat (1943–), had joined the team late in the summer. Wirth and Eizenstat wrote a note to the Secretary of State on September 4 outlining that the Byrd–Hagel Resolution made their current negotiating position "inadequate"[83] – an understatement if there ever was one. Making matters worse, the internal dissension among the Americans was well known by the delegates at Kyoto.[84] Between Berlin and Kyoto, the US strategy was in shambles, both at home and abroad, and Congress had further tied the hands of the negotiators with the Byrd–Hagel Resolution.

On December 1, 1997, the climactic moment had arrived as the meeting at Kyoto began (COP 3). The major players outlined their starting positions:[85]

- *The US* – pushed for the broadest mechanisms to achieve the emissions targets; reiterated their desire to see key developing countries participate, and they encouraged them to accept voluntary targets.
- *The EU* – wanted a 15 percent reduction in emissions compared to 1990, by 2000.
- *The G77* – the developed countries had to take the first steps. They supported the principle of "common but differentiated responsibilities," but were not willing to even discuss the "post-Kyoto evolutionary process."

Al Gore warned that his country was firm on the participation of key developing countries, but the Americans only asked the developing nations to reduce their rate of increase in emissions. The EU told the US to back off, stating that any discussions on new commitments for the developing countries were not helpful to the negotiations and were contrary to the Berlin Mandate.[86] On December 5, New Zealand put forward a compromise:[87]

> Annex I Parties' constituencies needed assurances that developing countries would adopt binding emissions limitation commitments in a third commitment period. He proposed double conditionality: Annex I Parties needed early agreement by non-Annex I countries on future commitments, but non-Annex I Parties would not be held to commitments if Annex I Parties did not fulfill their Kyoto commitments. He called for "progressive engagement" according to relative levels of development, and exemption for least developed countries. Supported by the US, CANADA, POLAND, SLOVENIA, AUSTRALIA, SWITZERLAND and JAPAN.

The G77 and China took the hard line that no commitments, even voluntary ones, would be put into the Protocol, and replied to the New Zealand proposal with a firm "no." Any remaining hope ended when the EU again waded in against the proposal, arguing that it violated the Berlin Mandate, and that further discussions on the principle of "common but differentiated responsibilities" would have to wait for future negotiations. At Kyoto, the defeated Americans stopped playing the game and simply ceded to the Protocol without gaining concessions from the developing countries.

All this begs the question: how did a group of the leading negotiators in the world end up in such a mess? In order to find the answer, we need to delve further into this story and consider the strategy used by the White House. Despite placating Congress with rhetoric about being committed to achieving emissions targets for developing nations, the White House pursued a separate agenda.

We can trace the failure of climate change policy back to the White House objective of joint implementation and the related flexible mechanisms, whereby an emissions-reduction project in a foreign country could be claimed by a nation within Annex I of the Protocol; in other words, a nation with emissions-reduction targets could choose the most cost-effective location for a project, even if in another country. It sounded like a good policy that industry would generally like and would improve the economics of climate change initiatives.

In the First Assessment Report of the IPCC in 1990, the concept of tradable emissions permits was extensively discussed in Chapter 9 on "Economic (Market) Measures" of the Working Group III Report, which stated:[88]

> 3) Economic instruments that may be relevant include: a system of tradable emission permits; emission charges; sanctions; subsidies to encourage the implementation of measures to limit/reduce emissions; and reduction of existing government interventions in other areas, such as transport, energy, food, and agriculture, that inadvertently encourage emissions of greenhouse gases. Indeed, an international system of tradeable emission permits or, alternatively, a system of international emission charges could offer the potential

of serving as a cost-efficient main instrument for achieving a defined target for the reduction of greenhouse gas emissions.

4) A number of contributors expressed major concerns about the use of tradeable emissions permits or sanctions. While it was agreed that studies should be continued on these instruments, contributors requested that their misgivings should be taken into consideration.

American interest in the program had begun during the first Bush Administration, which had suggested the use of "market-based incentives" in 1990:[89]

> To those who suggest we're only trying to balance economic growth and environmental protection, I say they miss the point. We are calling for an early new way of thinking to achieve both while compromising neither by applying the power of the marketplace in the service of the environment ... the United States is already taking action to stabilize and reduce emissions through our clean air legislation, our use of market-based incentives to control pollution.

And a related mechanism, joint implementation, had been a major policy initiative of the Clinton Administration's Climate Action Plan in 1993.[90] Several trial projects had already been approved, including a wind facility in Costa Rica and a solar project in Honduras.[91] Joint implementation became one of the three objectives of the American team at Berlin:[92]

1. first, to negotiate a mandate to begin considering next steps under the climate change convention
2. second, to ensure that all parties will be part of that process and advanced developing country's commitments
3. third, to establish a pilot phase for joint implementation.

Wirth placed special emphasis on the third objective: "We went into Berlin attempting to preserve our flexibility, which we have done, and to have the ability to implement joint implementation, which we have done."[93] This is why Wirth had declared victory in Berlin to the House Subcommittee. The US gained "advantages" from the Berlin Mandate related to the "enormous promise" of the pilot program on joint implementation.[94] Only Canada had initially supported the US position, but Wirth was proud of the fact that by the end of the meeting other nations were on board.[95] Initially, the EU officially had opposed the concept:[96]

> In addition, at the insistence of the US delegation led by then-Vice President Al Gore, emissions trading between countries was included as a flexible measure, together with the Clean Development Mechanism (CDM) and Joint Implementation. The European negotiating team felt that it had failed

to achieve most of what it had aimed for, and shortly after Kyoto most team members moved on to other assignments. Six months after Kyoto, new leadership at the Commission embraced emissions trading.

However, the American initiative provided a solution to some of the internal problems facing the EU. The original strategy had called for a harmonized carbon tax (proposed in 1992 but officially rejected in 1994), but a tax had to be approved by all the EU nations – and the UK,[97] especially, had strongly objected. Its failure to pass the carbon tax had been used by the US as an example of how the EU was hypocritical in its tough stance against climate change.[98] An environmental policy, such as the flexible mechanisms, was a much easier measure to adopt within the EU, as it only required a majority vote to pass.[99]

The importance of flexible mechanisms to the US provided a new opportunity for the EU to achieve its own objectives, so they were willing to cede[100] on the flexible mechanisms in exchange for the US agreeing to "adequate" targets.[101] By "adequate," the EU meant something close to their tough emissions targets, which was 15 percent below 1990 levels by 2010. A report by senior Clinton advisors called this target "naïve."[102] One may call the EU proposal many things, but naïve is not one of them, as the disorganized American negotiating team was being "held hostage"[103] on this issue by the Europeans.

At COP 2 (July 1996) in Geneva the US had agreed to legally binding emissions targets in exchange for the flexible mechanisms (including emissions trading),[104] but they did not seem to take the commitment seriously. On June 23, 1997, a special session was held by the UN, the "Rio Plus Five" (five years after the Rio conference), to review the progress of the Annex I countries. Observers noted the haggling between the EU and the US over legally binding targets.[105] By August, in a pre-Kyoto meeting in Bonn, most industrialized nations had submitted their emissions reduction target, but not the US.[106] As Kyoto approached, senior officials of the Clinton Administration were still struggling with the issue; some were suggesting ludicrous targets, such as returning to 1990 levels by 2040 (Janet Yellen, Chair of the Council of Economic Advisors), or even no targets at all (Larry Summers, Deputy Secretary of the Treasury).[107] By early October, still no target had been set.[108] The US announcement finally came a few weeks later, just a month before Kyoto; on October 22, President Clinton gave a speech to the National Geographic Society, where he listed three major points on climate policy:[109]

1.   the US would return its emissions to 1990 levels by 2008–12
2.   the US would promote the adoption of flexible mechanisms
3.   "developing countries must be engaged."

The EU would never let the US off so easily, as this was hardly the "adequate" target they had expected; an EU official called it "totally inadequate and downright irresponsible."[110]

At Kyoto, negotiations were at a stalemate until Al Gore arrived. In the final outcome, the US ended up with a target reduction of 7 percent below 1990 levels, which was similar to the final EU target of 8 percent. Even Bolin of the IPCC could not understand why these targets had been chosen, especially considering the objections of the US Senate.[111]

While the Americans may have argued that they succeeded in bringing down the EU target from 15 percent, the EU had only set such a high level as a negotiating ploy.[112] The higher the US target, the better the competitive advantage of the EU.[113] The US had ceded much – but they got their flexible mechanisms. The joint implementation appeared in Article 6 of the Kyoto Protocol and the principle of emissions trading in Article 17. Also, through a proposal from Brazil, a broader variation of the initial joint implementation was added with the Clean Development Mechanism (CDM).[114]

China and India also took advantage of the joint implementation so dear to the US, since the Americans had to agree to drop the calls for even voluntary emissions reductions by the developing countries.[115] Now, even this obscure hope of getting the developing nations in the game was gone. The flexible mechanisms were very costly to the Americans. Ironically, in the end, it was the EU and not the US which directly benefitted from the flexible mechanisms, as we will discuss in the next chapter. While the joint implementation/flexible mechanisms are excellent programs, their relevance to the US had been largely negated by the Byrd–Hagel Resolution, which made the participation of the developing countries mandatory for any chance of ratification of Kyoto by the Senate.

Why did the White House defy the Senate and agree to a treaty that could never be ratified? Why were the joint implementation and flexible mechanisms so important to the White House? The argument can be made that the flexible mechanisms would have been the only way for the Americans to meet their reduction targets,[116] but this point, even if valid, does not explain the concessions made by the Americans. While being a creative tool to combat climate change, the flexible mechanisms were primarily an administrative article and not a critical policy issue.

Who, then, designed the flawed strategy? Dingell, a senior Democrat at the Subcommittee hearings, was also looking for an answer to this question:[117]

Is the administration telling the State Department, what to do? Is the State Department hearing what the administration tells it to do, going off, doing its own thing? Is the administration perhaps being misled by the State Department, or indeed, am I perhaps being misled by the administration?

Tim Wirth, as the director of the portfolio, is an obvious suspect. He had been the political architect of "Project 88" for the US acid rain program (based on the Crocker–Dales model) which was the first major implementation of flexible mechanisms, and he stated that the program was "enormously important."[118] But it is unlikely that this was his own personal crusade, as Dingell had reported that the problems existed before Wirth became directly involved.[119] And, certainly, Wirth cannot be blamed for the outcome of Kyoto as he was not even there. The American objectives going into Kyoto with Eizenstat were virtually the same as the ones at Berlin with Wirth:

Berlin[120]	Kyoto[121]
First to negotiate a mandate to begin considering next steps under the climate change convention.	Realistic targets and timetables for reducing greenhouse gas emissions among the world's major industrial nations, which fully protect the unique role of our military in its global reach.
Second, to ensure that all parties will be part of that process and advanced developing country's commitments.	Meaningful participation of developing countries.
Third, to establish a pilot phase for joint implementation.	Flexible market-based mechanisms for achieving those targets cost-effectively.

When Eizenstat had gone to Kyoto he had been "instructed" by Bill Clinton and Al Gore that a priority was the flexible mechanisms.[122] Under the circumstances, one would have thought that the only priority going into Kyoto would have been to get the developing nations on board, as this had been demanded by the Byrd–Hagel Resolution. The House was thinking the same thing!

On March 4, 1998, a few months after Kyoto, the House of Representatives held another hearing (6th) by the Subcommittee on Energy and Power on "The Kyoto Protocol and Its Economic Implications," again chaired by Dan Schaefer. The Committee was in an angry mood. Criticism came from a written report by the Chairman of the Committee on Commerce, Tom Bliley:

> I am struck that the things that are claimed as victories from the negotiation appear to be battles not yet won, and the things the administration claims are still open to negotiation have in fact already been settled ... Prior to the Kyoto meeting, the President said that he was going to insist on "meaningful participation by developing countries." The testimony of both administration witnesses concede that the agreement does not require meaningful participation by developing countries ... how do we go there from here?

Eizenstat was as evasive as Wirth had been when answering the Subcommittee members, including the pointed questions from John Dingell – even sounding like Wirth when he called the flexible mechanisms "our concept, one little-known until Kyoto,"[123] the "greatest achievement at Kyoto"[124] and a "signal victory" by the US.[125] He reassured the Subcommittee that the Administration was firm on getting the commitment "to secure meaningful participation of key developing countries," and that they still had a chance of the developing countries volunteering to accept targets.[126] However, when pushed by the Committee, Eizenstat admitted that this option had already been excluded from the Protocol. Nevertheless, he vainly insisted that developing nations could be persuaded to adopt carbon targets.[127] Did Eizenstat really believe this malarkey, or was it his duty to defend the Administration to the bitter end?

Dingell could not contain himself about the misleading information that had been presented. In his opening reports, he chastised Eizenstat:[128]

> I had the pleasure of seeing Mr. Eizenstat over there in Kyoto, where I think we converted an opportunity for mischief into a genuine disaster.
>
> I know that you will try to defend an ill-conceived and badly executed Kyoto agreement on climate change.

The frustrated Representative from Michigan told Eizenstat that he had given the country a "very, very bad deal"[129] and could not be trusted. These were extremely harsh words indeed for a public committee meeting in Congress, especially from a member of the same party as the President.

After Kyoto, the US tried to initiate a new series of negotiations and, as a last option, at least get voluntary commitments from the developing nations.[130] This doomed plan only highlighted their desperate position. The Administration continued to lamely claim success, specifically the flexible mechanisms, about which they went on at great length.[131] Shortly before COP 4 (November 1998), the first meeting after Kyoto, Stuart Eizenstat summarized the "progress" that the US had made in climate change negotiations, commenting on the "spectacular diplomatic breakthroughs as we accomplished in Kyoto," such as the flexible mechanisms. He stated:[132]

> [O]ver the last year, we have also mounted a comprehensive diplomatic effort to encourage developing countries to become meaningfully involved in limiting the increase in their emissions, consistent with needed growth. While we intend to sign the Protocol to help lock in the gains made at Kyoto, the President has indicated he will not submit the Kyoto Protocol for advice and consent by the Senate without meaningful participation by key developing countries.

On November 12, 1998, Al Gore signed the Kyoto Protocol on behalf of the US government, but with a stipulation: "As we have said before, we will not submit the Protocol for ratification without the meaningful participation of key developing countries in efforts to address climate change."[133]

Many COP meetings would pass without the US coming any closer to ratifying the Kyoto Protocol. Negotiations continued, but any future proposals to discuss commitments by the developing countries were quickly squashed,[134] so there was no hope that the US would ratify the Protocol.[135] For the US, it was game over even before the Kyoto season had begun.

AmBush at the COP

On January 20, 2001, George W. Bush (1946–) took office. In terms of climate change, there was little difference between the policies of George W. and his father George H.W. The Bush doctrine on climate change was back, if it had ever left. Candidate Bush had presented his position on Kyoto in the Republican election platform:[136]

> Wherever it is environmentally responsible to do so, we will promote market-based programs that are voluntary, flexible, comprehensive, and cost-effective … Complex and contentious issues like global warming call for a far more realistic approach than that of the Kyoto Conference. Its deliberations were not based on the best science; its proposed agreements would be ineffective and unfair inasmuch as they do not apply to the developing world; and the current administration is still trying to implement it, without authority of law. More research is needed to understand both the cause and the impact of global warming. That is why the Kyoto treaty was repudiated in a lopsided, bipartisan Senate vote. A Republican president will work with businesses and with other nations to reduce harmful emissions through new technologies without compromising America's sovereignty or competitiveness, and without forcing Americans to walk to work.

And in an open letter to four Senators, including Chuck Hagel (1946–) of the Byrd–Hagel Resolution, Bush had reiterated his opposition to Kyoto:[137]

> As you know, I oppose the Kyoto Protocol because it exempts 80 percent of the world, including major population centers such as China and India, from compliance, and would cause serious harm to the U.S. economy. The Senate's vote, 95–0, shows that there is a clear consensus that the Kyoto Protocol is an unfair and ineffective means of addressing global climate change concerns.

On March 28, the new President officially announced that the US would not ratify the Kyoto Protocol. The reaction was a surprising world outcry,[138] as if the new Bush Administration was making a radical departure from US policy. Yet, Al Gore had said as much when he signed the Protocol. Bush's declaration was simply policy euthanasia, finally putting an end to the failed negotiations.

There were again allegations of the new Administration's abuse of science,[139] including:[140] censoring, suppressing and even dismissing federal scientists; altering, distorting and suppressing scientific findings for government reports; manipulating the government's science advisory system; and ignoring, distorting and selectively using scientific evidence in policy-making. Robert Watson, the Chairman of the IPCC, stated that "the only person who doesn't believe the science is President Bush."[141]

The Republican strategy to undermine the threat of climate change was outlined by the Washington lobbyist, Frank Luntz; his Republican playbook on environmental issues recommended that they should:[142]

1. assure their audience that you are committed to "preserving and protecting" the environment

2. provide specific examples on how the Federal Government is failing to protect the environment
3. discuss future plans
4. use the words "safer", "cleaner" and "healthier"
5. do not use business/economic jargon
6. state that they are seeking a "fair balance" between the environment and the economy
7. disclose the cost of environmental regulations
8. emphasize "common sense."

Luntz had a specific section on climate change, entitled "Winning the Global Warming Debate: An Overview," which listed five specific points:[143]

1. there is no consensus and there is a lack of certainty on the science
2. make the "right decision, not the quick decision"
3. encourage voluntary measures
4. emphasize the "fairness" of all nations participating, including China
5. Kyoto would hurt the economy, especially seniors and the poor.

An especially controversial action of the Administration took place in April 2002 at the Nineteenth Session of the IPCC during the election of the Bureau.[144] In an unprecedented move, the Bush Administration supported Pachauri of India against the incumbent Chairman, Watson of the US.[145] Many European nations, Brazil and South Africa, supported Watson, as did Bolin,[146] and he had officially been nominated by Portugal and the UK.[147] The American support of Pachauri was allegedly to promote a chairman from a developing country.[148] Many were suspicious of US motives, as Watson had been a Clinton appointee. About a week before the vote, concerns were publicly raised in the prestigious magazine, *Science*: "Senior researchers around the world fear that the U.S. move is part of a campaign to undermine the scientific credibility of IPCC."[149]

There may have been more to this story. According to newspaper reports, the government of George W. Bush had lobbied against Watson's re-election, after being encouraged to do so by ExxonMobil. The scandal was associated with a fax sent by Arthur G. (Randy) Randol, Senior Environmental Advisor of ExxonMobil, to John Howard,[150] Senior Associate Director of the Federal Council of Environmental Quality (CEQ).[151] The controversial fax stated that Randol would be calling Howard to discuss "the recommendations regarding the team that can better represent the Bush Administration interests" [on the IPCC]. The fax encouraged the Administration to remove all Clinton/Gore appointees to the IPCC, including Watson, as quickly as possible. Adding fuel to the fire were briefing papers given before meetings to the US Under-Secretary of State, Paul Dobriansky, that revealed:[152]

[T]he administration is found thanking Exxon executives for the company's "active involvement" in helping to determine climate change policy, and also seeking its advice on what climate change policies the company might find acceptable.

In another move to undermine the work of the IPCC, the Bush Administration created an open process for the government review process of the second draft of the AR4; the number of reviewer comments jumped to 90,000. The IAC criticized the Americans because it encouraged "orchestrated efforts by those with strong views about climate change to overwhelm the system."[153]

Misconduct penalty at Copenhagen

Copenhagen started off as the "Woodstock" of the COPs. At COP 15 in December 2009, over 40,000 people had applied to attend, and over 100 world leaders participated, including the new US President. Despite its own lack of carbon targets, a major diplomatic initiative was launched at Copenhagen by the Obama Administration to entice the developing countries to agree to new commitments. As Americans attempted to get off the bench and into the game, they ended up in the penalty box at Copenhagen.

The outcome was the controversial "Copenhagen Accord," which included developing nations.[154] Among the more than 110 countries accepting the Copenhagen Accord were the leading emerging nations:[155] China, India, Korea, South Africa, Brazil, Chile and Indonesia. At Copenhagen (COP 15), President Obama (1961–) stated:[156]

> Earlier this evening I had a meeting with the last four leaders I mentioned – from China, India, Brazil, and South Africa. And that's where we agreed to list our national actions and commitments, to provide information on the implementation of these actions through national communications, with international consultations and analysis under clearly defined guidelines. We agreed to set a mitigation target to limit warming to no more than 2 degrees Celsius, and importantly, to take action to meet this objective consistent with science ...
>
> Essentially you have a situation where the Kyoto Protocol and some of the subsequent accords called on the developed countries who were signatories to engage in some significant mitigation actions and also to help developing countries. And there were very few, if any, obligations on the part of the developing countries ...
>
> My view was that if we could begin to acknowledge that the emerging countries are going to have some responsibilities, but that those responsibilities are not exactly the same as the developed countries.

While the sound bites looked good in the press, the process almost caused the collapse of the COP. In their zeal to come up with the Copenhagen Accord, the Americans ruffled a lot of feathers – many of the members of the COP were excluded and basic protocols were bypassed, with most countries outraged more by the process than the actual agreement.[157] If nothing else, Copenhagen had witnessed the most intense involvement of world leaders in the climate change debate in more than a decade.[158] At COP 16 in Cancun, in December 2010, a few major procedural issues moved forward, but the most positive development was that delegates had been able to put the "debacle"[159] of Copenhagen behind them.

The Accord was not an official decision of COP 15, as the COP would only take "note." And the Executive Secretary of the UNFCCC, Yvo de Boer (1954–), stated: "We are not asking countries to adhere to the accord, we are asking them to associate with the accord."[160] It is disturbing that the head of the UNFCCC found it necessary to be so timid in his choice of words. Welcome to the pretty-please, maybe, world of the COP. While over 140 nations have signed on to the Copenhagen Accord, considering the lame wording, it is not clear that they have signed on to anything. De Boer resigned from his post six months later because the climate change negotiations were failing, using the following analogy: "You've got a bunch of international leaders sitting 85 stories up on the edge of a building saying to each other, you jump first and I'll follow. And there is understandably a reluctance to be the first one to jump."[161]

Playing solitaire – China climate change policy

Common but differentiated responsibilities

The failure of the US to join an international agreement on climate change can be assigned to the understanding of the short phrase "common but differentiated responsibilities." The original UNFCCC agreement at Rio acknowledged the statement and explained that developed countries "should take the lead in combating climate change"[162] and would carry a "disproportionate"[163] burden. For over 20 years, seasoned bureaucrats, especially those of America and China, have haggled over the meaning of these four simple words. In the view of China, the implications of this phrase were clear:[164]

> Developed countries should be responsible for their accumulative emissions during their 200-odd years of industrialization, which is the main reason for the current global warming, and they should naturally take the lead in shouldering the historical responsibilities to substantially reduce emissions. With regard to capabilities, developed countries have substantial economic strength and advanced low-carbon technologies, while developing countries lack the financial strength and technologies to address climate change, and face multiple arduous tasks of developing their economies, fighting poverty and addressing climate change. Therefore, developed countries should, on the one hand, take the lead in reducing emissions substantially, and, on the other, provide financial support and transfer technologies to developing countries. The developing countries, while developing their economies and fighting poverty, should actively adopt measures to adapt to and mitigate climate change in accordance with their actual situations.

Over the years, China had been steadfast in its position – there would be no emissions restrictions of any sort, even voluntary ones, for developing countries. A member of the Chinese delegation at Kyoto, when asked when they would agree to targets, replied: "Not this year. Not by 2010. Not by 2020. Not by 2050, and perhaps not ever."[165]

The dispute over responsibilities had been settled, more or less, at COP 1 in Berlin where only Annex I countries had carbon targets, but this did not end the debate.

The UN held a conference Rio+20 in June 2012 to renew political commitment to sustainable development. Members were asked for submissions for the event. The number of times the phrase "common but differentiated responsibility" was mentioned in the short submissions varied greatly among the participants:

US – 0[166]
EU – 1[167]
G77+China – 9[168]

China continues to place these four words front and center in the climate negotiations.

While the phrase "common but differentiated responsibility" is not without ambiguity, it is strange that it has been a stumbling block for so long. There is general agreement that differentiation was to be based upon a nation's present and historic emissions, and its economic ability to act on the problem of climate change; so let's examine how these criteria apply to China. Annual emissions of China at the time of Rio were significant, but still less than most of the developed world; however they were growing rapidly. Forecasts at the time of Kyoto had projected that China would be the largest emitter in the world by 2015.[169] In fact this actually happened about a decade earlier (Figure 3.1).

A fair comment made by China and others is that annual emissions had not caused the problem; it was historic (accumulated) emissions. At the time of Kyoto, these resided heavily with the developed nations, but the global picture has since changed; although China has only recently become a major contributor, it already has the second largest historic CO_2 emissions (the US has the most). Over the past half-century (1960–2010), the world's anthropogenic accumulated CO_2 emissions have been 1,000,000 Tg,[170] of which about 24 percent originated from the US and 12 percent from China. So both annual and historic emissions imply that China should have some carbon restrictions. Finally, with respect to economic ability, the most important of the qualifiers, China does not have the GDP per capita levels of the developed nations. But does this limitation mean that China should have no emissions restrictions?

From the early INC meetings and all the way up to Kyoto, there were numerous attempts to establish a definition of "differentiated" according to the economic status of the developing nations,[171] but these were repeatedly blocked by China.

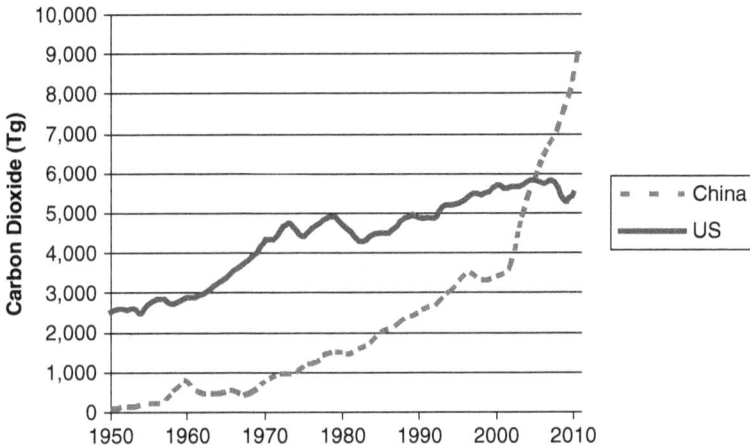

Figure 3.1. Annual CO_2 anthropogenic emissions from the US and China: 1950–2011. Source: CDIAC.

Why is China being so inflexible? There is a cultural distrust of the West and fears of "eco-colonialism."[172] Along with these suspicions, there is further support justifying China's hard-line stance:

1.   China has no legal responsibility to reduce emissions according to the Kyoto Protocol. If other nations expected China to have emissions targets, then why did they, including the US, agree to the Berlin Mandate?
2.   The developed nations have caused the problem. Were the developed nations being hypocritical asking the developing nations to do something?[173]
3.   The US was as adverse to binding targets as China. How could the Americans expect China to do what they have been unwilling to do?
4.   Exporting nations, where China is the leader, will naturally have greater emissions. Since most of these exports are destined for the US, why should China "pay" for emissions associated with US consumption?[174]
5.   Poverty is still widespread in China. Should China slow development and keep its people in poverty?

Let's look at each of these rationales in more detail. On the Kyoto obligations, the first point, the Chinese media has reminded the world that China has no internationally legal responsibilities on climate change.[175] At Berlin, at COP 1, Angela Merkel (1954–), then environment minister for Germany, had stated in her opening remarks:[176]

In accordance with the principles of the Convention, particularly in view of the common but differentiated responsibility, we, the industrialized countries, must be the first to prove that we are bearing our responsibility in protecting the global climate. Only when we demonstrate this by convincingly taking the lead, can we demand actions from other countries regarding climate protection.

While the Americans can claim that they did not *ratify* the Kyoto Protocol, they did *sign* the Protocol, which included the Berlin Mandate. How can they sign these international treaties which call for no emissions targets for China, and then demand that China accept such obligations?

As had been stated by China back at COP 1 in Berlin, the industrialized countries should meet their carbon targets before asking developing countries to do anything.[177] China did not cause the problem of climate change, so why should it help fix it? With this second rationale, China was among the more moderate of the developing countries (G77). In fact, many members of the developing nations were furious that the developed world had created a global crisis, and used the terms "historic responsibility," "guilt" and even "crimes" when discussing the emissions of the developed world.[178] Yet, when such terms were used at a meeting of the G77 at Kuala Lumpur in April 1992 (just before the second part of the fifth session of the INC), China discouraged such inflammatory statements and toned the rhetoric down.[179]

Fuelling China's hard-line position was the lack of any national legislation by the US. So, as to the third point, how could the US demand that China and other developing countries accept what they themselves had resisted throughout the INC and the COP process?[180] Twenty years later, the Americans still have not enacted any climate change legislation. Even if China agreed to legally binding carbon targets, could the US guarantee that it would do the same? Both the Clinton and Obama Administrations demonstrated that even with control of Congress, the President could not force climate change legislation through both Houses. China has actually done more to reduce the impact of climate change than the US and many members of the EU. At COP 17, in December 2011, the lead Chinese negotiator, Xie Zhenhua stated: "What qualifies you to tell us what to do? ... We are taking action. We want to see your action."[181] China is the only developing nation to have issued a policy on climate change[182] and certainly has outdone anything Washington has developed. On June 4, 2007, a new policy was issued to promote renewable energy, nuclear power and energy efficiency; and on September 22, 2009, the President, Hu Jintao (1942–), summarized China's commitment to climate change:[183]

We will endeavor to cut carbon dioxide emissions per unit of GDP by a notable margin by 2020 from the 2005 level. Secondly, we will endeavor to increase the share of non-fossil fuels in primary energy consumption to around 15% by 2020.

The disparaging raw data on emissions presents an inaccurate image of China's performance relative to the overall world community. The emissions of China are only at the world's average on a per capita basis. If China was emitting at the same per capita rate as Germany, which is often hailed as the leader in fighting climate change, their absolute emissions would have doubled! China is the global leader of renewable energy sources, in terms of investment,[184] capacity and development;[185] also, more of its electricity comes from renewable sources than the US or the EU.[186] In 2009, of the 80 gigawatts (GW) in new renewable energy capacity in the world, almost half (37 GW) was installed in China alone.[187] Overall, renewable energy capacity in China was 263 GW in 2010,[188] and in early 2011, China committed to reduce its carbon intensity by 40 percent by 2020 compared to 2005.[189]

The fourth rationale pertains to all the major exporting countries of the world, but especially China as it is the largest. Many nations have "deindustrialized" by closing their own facilities and sourcing from less expensive off-shore factories, thereby shipping their GHG emissions to others.[190] Exports account for about 25 percent of China's GHG emissions! Under current regulations, if a country increases its imports and decreases its production, its emissions decline. How does this help the fight against climate change when emissions are simply being shuffled from one region to another? Should China be held responsible for these "CO_2 exports"[191] on goods consumed in the United States? The carbon supply and demand issue opens an interesting discussion.[192] Which side of the carbon trade balance should be responsible for emissions?

Last, the prime reason for China's opposition to any outside restrictions on emissions rests on its need for economic development. Developing countries can ill afford the costs of fighting climate change.[193] China referred to the "luxury emissions" of the developed nations compared to the "survival emissions" of the developing world.[194] On June 4, 2007, the Chinese Assistant Foreign Minister, Cui Tiankai (1952–) stated: "We will abide by the Kyoto Protocol. But we believe that poverty alleviation must come first. Without poverty alleviation, environmental protection is just an empty phrase."[195]

A few years later Yu Qingtai, who had been the Chinese chief climate negotiator, made a similar comment.[196] As a leading emerging nation, China will be among the first to be labeled a developed nation. However, even at the rate China is growing, it has a long way to go before this happens.

Given these arguments, China's position has merit. In many ways, China has a strong case for its stance on climate change but, as we will show, there are more important considerations.

The road to Durban

The ongoing arguments over "common but differentiated responsibility" have produced negotiation fatigue. After COP 1 in Berlin, the Americans remained persistent in their efforts to somehow get China, and other emerging nations, to agree to emissions reductions. However, with the EU siding with China, the phrase "common but differentiated responsibility" only included administrative duties for developing countries and no requirements to reduce emissions or to even curtail the rate of emissions increases (Kyoto Protocol, Article 10). As China continued to "vociferously"[197] defend no restrictions on developing countries' emissions, the Americans had no chance of bringing down the Great Wall of China, reinforced as it was by the Berlin Wall.

After COP 4 in Buenos Aires, the issue of Kyoto targets for developing countries had been entirely off the agenda until COP 10 again at Buenos Aires in 2004,[198] when the delegates started to talk about post-2012 commitments and how to approach the sensitive issue of "common but differentiated responsibilities." Reluctantly, members agreed to have a "seminar" dealing with the contentious phrase in Bonn the following year. For the first time in almost a decade, the representatives were able to discuss this topic, at least in general terms.[199] At subsequent COP meetings, discussions continued:

- *COP 11* (December 2005, Montreal, Canada) – the role[200] of the developing countries moved forward. An outcome was the formation of the *Ad Hoc* Working Group on Further Commitments, but this negotiating body officially applied to developed countries only.
- *COP 12* (November 2006, Nairobi, Kenya) – Australia suggested at the AWG that all major emitters should be considered. China dismissed the idea.[201] Even voluntary reductions by developing countries were again rejected, and the proposal was deferred. The only positive aspect was that the controversial issue managed to stay on the negotiating table.
- *COP 13* (December 2007, Bali, Indonesia) – India suggested wording for "mitigation actions" by developing countries, which the US initially rejected and then reluctantly agreed to.[202] The "Bali roadmap" was an opportunity to get the climate change negotiations on track, not just with the inclusion of the developing nations but also the United States.
- *COP 14* (December 2008, Poznan, Poland) – this was an intermediate meeting on the way to Copenhagen. A suggestion to define different categories of developing countries was once more rejected.[203]
- *COP 15* (December 2009, Copenhagen, Denmark) – the controversial "Copenhagen Accord" was reached, as discussed earlier. Afterwards in October 2010, there was a climate conference in China in preparation for the COP in Cancun. Frustrated US delegates accused their Chinese hosts of reneging on their agreement in Copenhagen.[204]
- *COP 16* (December 2010, Cancun, Mexico) – there was debate over whether there should be a second commitment period to Kyoto or a new treaty negotiated.[205]

The head of the Chinese Climate Change Office of the National Development and Reform Commission, Su Wei, added in late 2010: "Right now there are still huge differences between developed and developing countries in the negotiations on climate change problems."[206] He went on to state that the major obstacle remained "common but differentiated responsibilities," and two weeks before the Cancun Conference (COP 16), Ambassador Song Zhe addressed the European Parliament where he stressed "the great importance of the principle of *common, but differentiated responsibilities.*"[207]

The world was especially watching China at COP 17 in December 2011 in South Africa, where the press hailed China as the "rock star at Durban."[208] Xie Zhenhua, the lead Chinese negotiator, did not disappoint, as he opened the door a crack to potential Chinese participation. The contentious principle of "common but differentiated responsibilities" was raised early in the meeting by the Chinese delegation.[209] The EU, for once, talked tough and demanded that China and other emerging economies would have to directly participate. At the end of the conference delegates agreed to the Durban Platform, which included plans for future negotiations for a post-2020 agreement that would involve all nations, including China, India and the United States. The Environment Minister of Poland, Marcin Korolec, declared: "This is a moment comparable only to, if not surpassing, the COP1 from 1995, when the Berlin mandate was established which led to the Kyoto protocol."[210]

The American view was more down to earth, "the sales job just went from impossible to very hard,"[211] and Secretary of State Hillary Clinton declared:[212]

> But coming out of Durban, just now, you know, the Chinese were dead set against accepting responsibilities that were in any way comparable to what they still referred to as developed countries … Our position is, you are now the largest greenhouse gas emitter in the world; we cannot act as though you are Botswana, I mean – or the Seychelles. I mean, you have to take responsibility.
>
> The deal that was hammered out – by no means perfect, before any expert in, you know, climate claims otherwise here – was to, for the first time, end this differentiation between the developed and the developing in terms of what we all have to do to meet this global challenge.

Xie Zhenhua's comments at the end of COP 17 also reminded everyone that the game would not be won easily:[213]

> The outcome is fully in accordance with the mandate of the UN Climate Change Convention and its Kyoto Protocol and the Bali Roadmap, the two-track negotiation process and the principle of common but differentiated responsibilities … What needs to be pointed out is that developed countries lack the political will to reduce emissions and provide finance and the transfer of technology to support developing countries, which is a major obstacle for international cooperation in addressing climate change.

And negotiations on the details of the Durban Platform did not begin well in May 2012:[214]

> However, negotiations in 2012 got off to an inauspicious start and the Bonn Climate Change Conference was marred by mistrust and unabashed posturing. The meeting was almost paralyzed by prolonged procedural wrangling, which many described as "unprecedented" … Many could not begin to imagine how difficult it would be to begin implementing the Durban decisions.

As phase one of the Kyoto season ended at Doha (COP 18), the process continued, but just barely. Among the issues for discussion were the infamous "common but differentiated responsibilities." [215]

The World Series of the carbon game

In the World Series of the carbon game, the Yankees had been the perennial winners, but since 2006 the Tigers have taken the crown. The tiger economy of China has led this nation to be the largest CO_2 emitter in the world.[216] Coal is the major domestic resource available as fuel for electricity generation. The energy content of coal produced and consumed in China was equivalent to all the oil produced by OPEC in 2010![217] With its turbo-charged economy, and unprecedented coal-fired electricity capacity expansion, CO_2 emissions have gone wild. Despite much hype about the Kyoto Protocol and new renewable energy projects, global GHG emissions have increased between 2000 and 2009 far more than in any other decade in history, and over 70 percent of this increase has come from China.[218] In 2011 the increase in annual emissions from China was the same as the entire annual emissions of Germany, the sixth largest emitter in the world.

China has been playing a carbon game of solitaire, choosing to be the dealer of its own hand; however, China is an essential participant in the fight against climate change, and the international community cannot play the carbon game without it. The refusal to adopt carbon targets by China and other developing countries, however legitimate, creates a practical problem. If the entire world achieved on a per capita basis the "American dream," then global emissions would rise from 30,000 Tg to 130,000 Tg per annum. Without some restrictions on the growth of emissions in developing countries, it will be impossible to bring global emissions under control. One can make some good arguments to justify why China should not have to participate, yet these arguments do not help the planet as China's insatiable appetite for energy has transformed a domestic problem into an international crisis. To illustrate this point, if the US and the EU carbon emissions had been zero in 2010, the world emissions still would have been about the same as 1990, the Kyoto base year! Even if they tried their best, the US and the EU cannot stop climate change without China.

A minor league team – Canada climate change policy

Canada[219] can be viewed as a microcosm for the typical factors influencing political positions on climate change. Three provinces in particular, Quebec, Ontario and Alberta, which have the largest provincial emissions in the country (Table 3.1), underscore the conflicts and contradictions that arise when dealing with climate change. The range of their support covers the entire spectrum, with Alberta at one extreme and Quebec at the other, and with Ontario somewhere in between blazing down a path that no major region in the world has dared to take before.

Alberta has become a favorite climate change kicking horse by some provinces; yet, Canadians do not mind the enormous contribution that its energy industry has

Table 3.1. Carbon statistics (annual emissions) for the top five provinces in Canada (expressed as Tg CO_{2e}).

	1990	1995	2000	2005	2010
Canada	589	641	718	740	692
Quebec	84	79	85	86	82
Ontario	176	174	206	206	171
Saskatchewan	43	59	65	70	72
Alberta	166	200	219	228	233
British Columbia	49	58	62	63	56

Source: Environment Canada, Part III, pp. 13, 59, 61, 65, 67, 69 and earlier reports for data from 1995.

made to the country's economy. Increasing oil production has come from the oil sands, which mainly goes to the US. Compounding their carbon problem, most of the province's electrical requirements are supplied by coal-fired utilities. The coal comes from Alberta coal mines which allows the province to have some of the lowest energy costs in the world. Many of the province's coal-fired utilities have delivered fuel costs of $1 per GJ (gigajoule). Alberta has the perfect conditions to ensure it is among the worst offenders in the world as it would incur major costs to reduce its emissions, and the provincial economy could be harmed. Not surprisingly, then, the Alberta government has been a vocal opponent of Kyoto,[220] and the leader of the opposition Wildrose Alliance Party, Danielle Smith, is even a climate skeptic.[221] Nevertheless, it was the Alberta government that passed the first climate change legislation in North America on July 1, 2007 (Albertan Climate Change and Emissions Management Act).[222] While we would not expect an oil-producing region to enact such legislation, Alberta is an exception, showing that climate change legislation can move forward in a fossil fuel dominated economy.

At the other end of the spectrum are Quebec and British Columbia, which are lucky enough to have most of their electricity come from hydro power. The two provincial governments are the strongest advocates of strict and binding GHG reductions in the country. For instance, despite divisive politics within Quebec, all three of its political parties unanimously endorsed Kyoto.[223] Does this mean that Quebec is more environmentally conscious than Alberta, or, rather, that it stands to make money off the Kyoto targets from its vast hydro resource base?

Between Alberta and Quebec, geographically and in terms of energy mix when it comes to electricity generation, sits Ontario, which has a mishmash of everything – coal, hydro and nuclear. What really sets Ontario apart from the rest of Canada, and the world, is its commitment to replace all of its coal for electricity production by the end of 2014.[224] Between 2005 and 2011, coal use had already been reduced by 87 percent.[225] The elimination of coal-based electricity generation will reduce GHG emissions by 61 Tg of CO_{2e} per year (similar to the annual emissions of Portugal). This bold action by Ontario represents one of the most meaningful GHG initiatives in the world.

During the 1980s and 1990s, Canada had played a major role in climate change, the UNFCCC, INC, COP and IPCC. But during the last decade of the twentieth

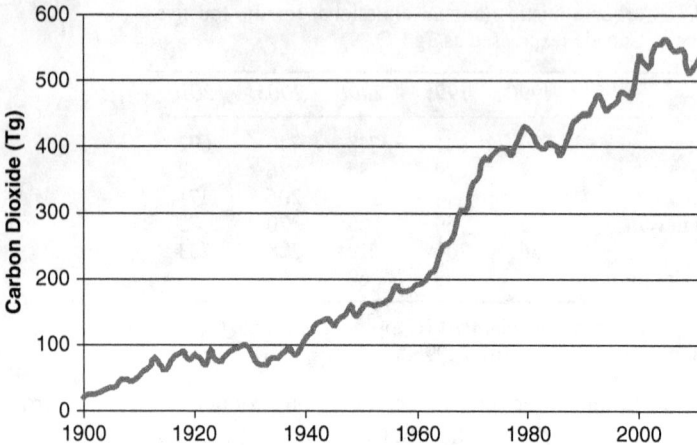

Figure 3.2. Annual CO_2 anthropogenic emissions in Canada: 1900–2011.
Source: CDIAC.

century something happened. Canadian emissions had grown steadily and by 2002 had risen by 16 percent (Figure 3.2), compared to the Kyoto base year of 1990, but the country was supposed to actually reduce its emissions by 6 percent. The per capita GHG rating also places Canada among the highest (i.e., worst) in the world. Canada's carbon footprint makes it the Sasquatch of carbon emitters in the world.

Both Canada and Germany had ratified the Kyoto Protocol in 2002. The Canadian target was a 6 percent reduction in emissions compared to 1990, while Germany's was a reduction of 21 percent. The carbon targets speak for themselves, or do they? Germany will likely meet its Kyoto target while Canada's emissions will not even be close. What has Canada done wrong? While there is no excuse for Canada ratifying the Kyoto Protocol, the carbon targets in the treaty were unfair. Germany's emissions were decreasing during the 1990s as Canada's were increasing. By 2002, when ratification took place, Germany's emissions were already below 1990 levels by 18 percent; these had been driven only by economic factors. Canada on the other hand was already over 1990 levels by 16 percent. So in the year of ratification of the Kyoto Protocol a very different picture of the carbon targets had emerged; Canada was 22 percent away from its target and Germany only 3 percent. Canada's global competitive position would have been seriously harmed had it tried to meet its targets, but there was also a penalty for not reaching the Kyoto target – the purchase of up to $14 billion in carbon credits.[226] The government had little choice and, on December 12, 2011, Canada became the only nation to officially withdraw from the Kyoto Protocol. Some COP members objected to Canada's gift of coal for Christmas. However, the Kyoto Protocol had placed a costly burden on North America and offered Germany and other European nations a free ride.

References[227]

Aden, N., Fridley, D., and Zheng, N., *China's Coal: Demand, Constraints, and Externalities*, Berkeley National Laboratory, July 2009; http://china.lbl.gov/publications/china%E2%80%99s-coal-demand-constraints-and-externalities

Blackwell, B., De Boer says Copenhagen Accord deadline is 'soft', *Recharge*, p. 3, January 22, 2010

Bodansky, D., The Copenhagen Climate Change Conference: A Post-Mortem, February 12, 2010; http://papers.ssrn.com/sol3/papers.cfm?abstract_id=1553167

Bolin, B., *A History of the Science and Politics of Climate Change*, Cambridge University Press, 2007

Bolin B., Politics and the IPCC, *Science* 296 (5571), 1235, 2002

BP, *Statistical Review of World Energy*, 2013 (and earlier years); http://www.bp.com/en/global/corporate/about-bp/statistical-review-of-world-energy-2013.html

Broder, J.M., Climate Talks in Durban Yield Limited Agreement, *New York Times*, December 11, 2011; http://www.nytimes.com/2011/12/12/science/earth/countries-at-un-conference-agree-to-draft-new-emissions-treaty.html?_r=1&scp=2&sq=durban&st=cse

Buckley, C., China Says It is World's Top Greenhouse Gas Emitter, *Reuters*, November 23, 2010; http://www.reuters.com/article/idUSTRE6AM1NG20101123?pageNumber=1

Buckley, C., China Vows to Cut Energy, Carbon Intensity by 2015, *PlanetArk*, February 28, 2011; http://planetark.org/wen/61334

Bush George, Presidential Library and Museum, *Remarks at the Closing Session of the White House Conference on Science and Economics Research Related to Global Change*, April 1990; http://bushlibrary.tamu.edu/research/public_papers.php?id=1772&year=1990&month=4

Cass, L.R., *The Failures of American and European Climate Policy*, SUNY, 2006

CBC News, Canada Pulls Out of Kyoto Protocol, December 12, 2011; http://www.cbc.ca/news/politics/story/2011/12/12/pol-kent-kyoto-pullout.html

CBC News, Quebec to Stand Up for Kyoto Even if Ottawa Won't, November 28, 2007; http://www.cbc.ca/canada/montreal/story/2007/11/28/qc-kyoto1128.html

CDIAC (Carbon Dioxide Information Analysis Center) – Oak Ridge National Laboratories – US Department of Energy; data taken on May 12, 2013; (note that historic data is periodically updated by CDIAC); http://cdiac.ornl.gov/trends/emis/meth_reg.html

China Climate Change Info-net, *China's Policies and Actions for Addressing Climate Change*, 2012; http://www.ccchina.gov.cn/WebSite/CCChina/UpFile/File1324.pdf

China Daily, Carbon Trading in Pipeline, July 22, 2010; http://www.chinadaily.com.cn/china/2010-07/22/content_11033249.htm

China Electricity Council, *News, Huadian Power International Buys Into Upstream Businesses*; http://english.cec.org.cn/htmfiles/2009717132838.html

Clinton, H., Secretary Clinton: Democracies Must Have 'Habits of the Heart', PBS Newshour, December 14, 2011; http://www.pbs.org/newshour/bb/world/july-dec11/clinton_12-14.html

Cole, D.H., A Glimmer of Hope, The EU's Emissions Trading Scheme, Chapter 3, 2009; http://papers.ssrn.com/sol3/papers.cfm?abstract_id=1533566

Convery, F., Ellerman, D., and de Pertius, C., The European Carbon Market in Action: Lessons from the First Trading Period, March 2008; http://web.mit.edu/globalchange/www/ECMreport.html

Davidson, O.G., Can the U.S. Create its Own Energy Revolution, *Inside Climate News*, November 20, 2012; http://insideclimatenews.org/news/20121120/germany-energiewende-clean-energy-economy-solar-wind-electric-grid-utilities?page=show; accessed November 2012

De Souza, M., It's Official: Harper Government Withdraws from Kyoto Climate Agreement, Megan Leslie, Member of Parliament, December 14, 2012; http://meganleslie.ndp.ca/post/its-official-harper-government-withdraws-from-kyoto-climate-agreement

Earth Negotiations Bulletin; http://www.iisd.ca/vol12/

Eizenstat, S., Fighting Global Warming: From Kyoto to Buenos Aires and Beyond, Speech at the Center for National Policy, Washington, DC, October 28, 1998; http://www.mtholyoke.edu/acad/intrel/stuart2.htm

Embassy of the People's Republic of China in Canada, *China Firm on Track of Addressing Climate Change*, December 5, 2011; http://ca.china-embassy.org/eng/dsxx/dsjh/t884320.htm

Environment Canada, *National Inventory Report, Greenhouse Gas Sources and Sinks in Canada: 1990–2010*; http://unfccc.int/national_reports/annex_i_ghg_inventories/national_inventories_submissions/items/6598.php

European Commission, *A Community Strategy to Limit Carbon Dioxide Emissions and to Improve Energy Efficiency*, Sec. (91) 1744, October 1991; http://aei.pitt.edu/4931/

Feinberg, S., Clean Energy Investment Storms to New Record in 2010, *Bloomberg New Energy Finance*, January 11, 2011; http://bnef.com/Download/pressreleases/134/pdffile/

Gelb, L.H., FOREIGN AFFAIRS; Sununu vs. Scientists, *New York Times*, February 10, 1991; http://www.nytimes.com/1991/02/10/opinion/foreign-affairs-sununu-vs-scientists.html?scp=1&sq=sununu%20vs.%20scientists&st=cse

Gore, A., The Selling of an Energy Policy, *New York Times*, April 21, 2002; http://www.nytimes.com/2002/04/21/opinion/21GORE.html

Grubb, M., Brewer, T.L., Sato, M., Heilmayr, R., and Fazekas, D., Climate Policy and Industrial Competitiveness: Ten Insights from Europe on the EU Emissions Trading System, August 2009; http://www.climatestrategies.org/component/reports/category/61/204.html

Grubb, M., Hourcade, J.-C., and Oberthur, S., Keeping Kyoto, Climate Strategies, July 2001; http://www.climatestrategies.org/research/our-reports/category/58/190.html

The Guardian, China Defends Carbon Emissions Growth, November 22, 2012; http://www.guardian.co.uk/environment/2012/nov/22/china-defends-carbon-emissions-growth

Harrison, K., The Road Not Taken: Climate Change Policy in Canada and the United States, *Global Environmental Politics*, 7:4, p. 99, November, 2007; http://www.politics.ubc.ca/fileadmin/template/main/images/departments/poli_sci/Faculty/harrison/Canada_US_august.pdf

Harvey, F., Climate Change Policy: Cancun Offers Slim Hope of Progress, *Financial Times*, October 29, 2010; http://www.ft.com/cms/s/0/6fb5f38a-e2e8-11df-9735-00144feabdc0.html#axzz29TAuFJHK

Hecht, M.T., and Tirpak, D., Framework Agreement on Climate Change: A Scientific and Policy History, *Climatic Change* 29 (4), 371, 1995.

Helm, D., Climate Change Policy: Why Has So Little Been Achieved; *Oxford Review of Economic Policy*, 24 (20), 211, 2008; http://www.kysq.org/docs/211.pdf

Helm, D., Forget Kyoto: Putting a Tax on Carbon Consumption, *environment360*, November 8, 2012; http://e360.yale.edu/feature/forget_kyoto_putting_a_tax_on_carbon_consumption/2590/#.UJvppSSSrdI.twitter

Helm, D., To Slow Warming, Carbon Tax, *New York Times*, November 11, 2012; http://www.nytimes.com/2012/11/12/opinion/on-climate-change-the-us-is-doing-better-than-europe.html?partner=rssnyt&emc=rss&_r=0; accessed November 2012

Historycentral.com, *2000 Republican Platform*; http://www.historycentral.com/elections/Conventions/RepPlat2000.html

Hu Jintao, Join Hands to Address Climate Challenge, *United Nations Summit on Climate Change*, September 2009; http://www.un.org/wcm/webdav/site/climatechange/shared/Documents/China.pdf

IEA, Boxer to Focus on Climate Science, *Energy Efficiency and Climate Change News*, December 8–14, 2010; http://www.iea.org/newsletters/ee_cc/index.asp; accessed November 2011

InterAcademy Council, *Climate Change Assessments, Review of the Processes and Procedures of the IPCC*, 2010; http://www.interacademycouncil.net/24026/26050.aspx

IPCC, *First Assessment Report, Working Group III: The IPCC Response Strategies*, 1990

IPCC, *Report of the Third Session of WMO/UNEP Intergovernmental Panel on Climate Change*, Washington, DC, February 5–7, 1990

IPCC, *Report of the Fifth Session of WMO/UNEP Intergovernmental Panel on Climate Change*, Geneva, March 13–15, 1991

IPCC, *Report of the Nineteenth Session of the Intergovernmental Panel on Climate Change*, Geneva, April 17–20, 2002

Jehl, D., *New York Times*, April 4, 2001; http://www.nytimes.com/2001/04/04/us/us-rebuffs-european-plea-not-to-abandon-climate-pact.html?scp=1&sq=&st=nyt

Lan, L., and Jing, L., UN Climate Deal Salvaged in Durban, *ChinaDaily USA*, December 12, 2011; http://www.chinadaily.com.cn/usa/china/2011–12/12/content_14251079.htm; accessed December 2011

Lawyler, A., Battle Over IPCC Chair Renew's Debate on U.S. Climate Policy, *Science* 296 (5566), p. 232, 2002

Lean, G., Melting Himalayan Glaciers: No Peaceful End to the Scandal, *The Telegraph*, January 22, 2010; http://www.telegraph.co.uk/earth/earthcomment/geoffrey-lean/7055303/Melting-Himalayan-glaciers-no-peaceful-end-to-the-scandal.html

Lim, L., China's Coal Fueled-Boom has Costs, *NPR News*, May 2, 2007; http://www.npr.org/templates/story/story.php?storyId=9947668&ft=1&f=1007&sc=emaf&sc=emaf

Lou, Y., and Lawrence, W., China to Cut Greenhouse Emissions by 950 Million Tons (Update 1), *Bloomberg*, June 4, 2007; http://www.bloomberg.com/apps/news?pid=2060 1103&sid=aM8H5XjNVbFQ

Luntz, F., *Luntz Memo on the Environment*, cf. Environmental Working Group; http://archive.is/RXOpO; March 2003

Mackenzie, D., Making Things the Same: Gases, Emission Rights, and the Politics of Carbon Markets, *Insights* 1 (6), 2008; http://www.dur.ac.uk/resources/ias/Mackenzie15Aug.pdf

Mackenzie, D., Too Hot for Head of Climate Panel, *New Scientist* 174 (2399), 16, 2002

Mason, G., Alberta's Wildrose Leader is no Shrinking Violet, *Globe & Mail*, December 16, 2009; http://www.theglobeandmail.com/news/opinions/albertas-wildrose-leader-is-no-shrinking-violet/article1403149/

Max, A., Ex-Climate Chief: Talks are Failing, *The Times-Tribune*, December 5, 2011; http://thetimes-tribune.com/news/health-science/ex-climate-chief-talks-are-failing-1.1240777#axzz1ffslHTuk

McCright, A.M., and Dunlap, R.E., Anti-reflexivity: The American Conservative Movement's Success in Undermining Climate Science and Policy, *Theory, Culture & Society* 27 (2–3), 100, 2010

Ministry of Foreign Affairs of the People's Republic of China, *Remarks by H.E. Ambassador Song Zhe at the Roundtable Discussion with MEPs on Climate Change*, November 10, 2010; http://www.fmprc.gov.cn/eng/wjb/zwjg/zwbd/t768292.htm

Mintzer, I.M., and Leonard, J.A., (eds.), *Negotiating Climate Change*, Cambridge University Press, 1994.

Mueller, B., Michaelowa, A., and Vrolijk, C., Rejecting Kyoto; Climate Strategies, http://www.climatestrategies.org/component/reports/category/58/192.html

Murse, T., A Brief History of White House Solar Panels, About.com, US Government Info, n.d.; http://usgovinfo.about.com/od/thepresidentandcabinet/tp/History-of-White-House-Solar-Panels.htm

National Research Council, National Academy of Sciences, *Carbon Dioxide and Climate: A Scientific Assessment*, July 1979; http://www.nap.edu/openbook.php?record_id=12181

New York Times, Some White House Effect, April 21, 1990; http://www.nytimes.com/1990/04/21/opinion/some-white-house-effect.html

Ontario, Ontario Getting Out of Coal-fired Generation, *Newsroom*, January 9, 2013; http://news.ontario.ca/mei/en/2013/1/ontario-getting-out-of-coal-fired-generation.html

Ontario, Ontario's Coal Phase Out Plan, *Newsroom*, September 3, 2009; http://www.news.ontario.ca/mei/en/2009/09/ontarios-coal-phase-out-plan.html

Pearce, F., The Week the Climate Changed, *New Scientist*, 52, October 15, 2005

Pooley, E., *The Climate War*, Hyperion, 2010.

Posner, E.A., and Sunstein, C.R., Climate Change Justice, August 2007; http://papers.ssrn.com/sol3/papers.cfm?abstract_id=1008958

Randol, A.G., Fax to John Howard, *Regarding: Bush Team for IPCC Negotiations*, February 6, 2001; http://www.nrdc.org/media/docs/020403.pdf

Ren21, *Global Status Report*; http://www.ren21.net/REN21Activities/Publications/GlobalStatusReport/tabid/5434/Default.aspx

Renewables International, Can Renewables Replace Coal, January 11, 2013; http://www.renewablesinternational.net/can-renewables-replace-coal/150/537/59611/

Revkin, A.C., Bush vs. the Laureates: How Science Became a Partisan Issue, *New York Times*, October 19, 2004; http://www.nytimes.com/2004/10/19/science/19poli.html

Revkin, A.C., China Sustains Blunt 'You First' Message on CO_2, *New York Times*, September 2, 2010; http://dotearth.blogs.nytimes.com/2010/09/02/china-sustains-blunt-you-first-message-on-co2/

Rio+20, United Nations Conference on Sustainable Development, European Union and its Member States, November 1, 2011; http://www.uncsd2012.org/rio20/index.php?page=view&type=510&nr=240&menu=20

Rio+20, United Nations Conference on Sustainable Development, Group of 77 and China, November 1, 2011; http://www.uncsd2012.org/rio20/index.php?page=view&type=510&nr=399&menu=20

Rio+20, United Nations Conference on Sustainable Development, United States of America, November 1, 2011; http://www.uncsd2012.org/rio20/index.php?page=view&type=510&nr=370&menu=20

Schneider, S., *Science as a Contact Sport, Inside the Battle to Save Earth's Climate*, National Geographic Society, 2009

Shabecoff, P., Global Warming has Begun, Expert Tells State. *New York Times*, June 24, 1988; http://www.nytimes.com/1988/06/24/us/global-warming-has-begun-expert-tells-senate.html?pagewanted=1

Shabecoff, P., Joint Effort Urged to Guard Climate, *New York Times*, January 31, 1989; http://www.nytimes.com/1989/01/31/science/joint-effort-urged-to-guard-climate.html?ref=philip_shabecoff

Simpson, J., Jaccard, M., and Rivers, N., *Hot Air, Meeting Canada's Climate Change Challenge*, Emblem, 2007.

Stanway, D., China Says Rich-Poor Divide Still Dogs Climate Pact Talks, *Reuters*, September 13, 2010; http://uk.reuters.com/article/idUKTRE68C0RS20100913

Stevens, W.K., Nations to Consider Toughing Curbs on Global Warming, *New York Times*, February 21, 1995; http://www.nytimes.com/1995/02/21/science/nations-to-consider-toughening-curbs-on-global-warming.html?scp=1&sq=Nations+to+consider&st=nyt

The Climate Change Action Plan; http://www.gcrio.org/USCCAP/index.html

UNFCCC, *Conference of the Parties, Second Session – Part I*, Geneva, October 29, 1996; http://unfccc.int/resource/docs/cop2/15.pdf

UNFCCC, *Conference of the Parties, Fifteenth Session – Draft Decision*, Copenhagen, December 18, 2009; http://unfccc.int/resource/docs/2009/cop15/eng/l07.pdf

UNFCCC, *Conference of the Parties, Fifteenth Session – Part II*, Copenhagen, March 30, 2010; http://unfccc.int/resource/docs/2009/cop15/eng/11a01.pdf

UNFCCC, *United Nations Framework Convention on Climate Change*, 1992; http://unfccc.int/resource/docs/convkp/conveng.pdf

United Nations, *Intergovernmental Negotiating Committee for a Framework Convention on Climate Change, First Session*, New York, March 8, 1991; http://unfccc.int/resource/docs/a/06.pdf

United Nations, *Intergovernmental Negotiating Committee for a Framework Convention on Climate Change, Second Session, Item 2 of the Provisional Agenda*, May 22, 1991; http://unfccc.int/resource/docs/1991/a/eng/misc01a01.pdf

United Nations, *Intergovernmental Negotiating Committee for a Framework Convention on Climate Change, Third Session, Compilation of Proposals Related to Commitments*, August 15, 1991; http://unfccc.int/resource/docs/1991/a/eng/misc07.pdf

US Department of State, *Climate Change: An Emerging Policy on Developing Country Participation, Information Memorandum*, September 4, 1997; http://www2.gwu.edu/~nsarchiv/NSAEBB/NSAEBB303/Document%206.pdf

US Department of State, *Climate Change Scenarios, Information Memorandum*, September 15, 1997; http://www2.gwu.edu/~nsarchiv/NSAEBB/NSAEBB303/Document%208.pdf

US Department of State, OES, *Climate Change Negotiations: Key Issues Facing US in the Run-up to the Kyoto Conference*, September 5, 1997; http://www2.gwu.edu/~nsarchiv/NSAEBB/NSAEBB303/Document%205.pdf

US Department of State, *State of Play on Climate Change, Information Memorandum*, October 2, 1997; http://www2.gwu.edu/~nsarchiv/NSAEBB/NSAEBB303/Document%2014.pdf

US Department of State, *The Kyoto Meeting: A Scenario for Redefining Success*, September 24, 1997; http://www2.gwu.edu/~nsarchiv/NSAEBB/NSAEBB303/Document%2012.pdf

US House of Representatives, Hearings before the Subcommittee on Energy and Power, 104th Congress, Serial Nos. 104–13, *International Global Climate Change Negotiations*, March 21 and May 19, 1995

US House of Representatives, Hearing before the Subcommittee on Energy and Power, 105th Congress, Serial Nos. 105–108, *The Kyoto Protocol and Its Economic Implications*, March 4, 1998; http://books.google.ca/books?id=q19o4MUr87cC&pg=PA13&dq=Stuart+Eizenstat,+kyoto&hl=en&ei=27NqTMnCJY30tgOE2rgw&sa=X&oi=book_result&ct=result&resnum=1&ved=0CCoQ6AEwAA#v=onepage&q=Stuart%20Eizenstat%2C%20kyoto&f=false

US Senate, *Byrd–Hagel Resolution*, 105th Congress, 1st Session, S. Res. 98, July 21, 1997, National Center for Public Policy Research; http://www.nationalcenter.org/KyotoSenate.html

Usui, Y., New Modes of Governance and the Climate Change Strategy in the European Union, 2005; http://project.iss.u-tokyo.ac.jp/crep/pdf/ws05/3pa.pdf

Van Renssen, S., Durban Leaves Large Question Marks Hanging Over Climate Policy, *European Energy Review*, December 15, 2011; http://www.europeanenergyreview.eu/site/pagina.php?id_mailing=238&toegang=ac1dd209cbcc5e5d1c6e28598e8cbbe8&id=3425

Vidal, J., Revealed: How Oil Giant Influenced Bush, *Guardian*, June 8, 2005; http://www.guardian.co.uk/news/2005/jun/08/usnews.climatechange

Vidal, J., If Rajendra Pachauri Goes, Who on Earth Would Want to be IPCC Chair?, *Guardian*, September 3, 2010; http://www.guardian.co.uk/environment/cif-green/2010/sep/03/rajendra-pachauri-ipcc

Warrick, J., U.N. Summit Reaches an Ineffectual End, *Washington Post*, June 28, 1997; http://www.highbeam.com/doc/1P2-722524.html

Weart, B., *The Discovery of Global Warming, Roger Revelle's Discovery*, Harvard University Press, 2008; http://www.aip.org/history/climate/Revelle.htm

Weekly Compilation of Presidential Documents, *Remarks at the National Geographic Society*, October 22, 1997; Vol. 33, No. 43 (Monday, October 27, 1997), pp. 1629–34; http://www.gpo.gov/fdsys/pkg/WCPD-1997-10-27/html/WCPD-1997-10-27-Pg1629.htm

White House, President Barack Obama, *Remarks by the President During Press Availability at Copenhagen*, December 18, 2009; http://www.whitehouse.gov/the-press-office/remarks-president-during-press-availability-copenhagen

White House, President George W, Bush, *Interview of the President by ITV*, United Kingdom, June 29, 2005; http://georgewbush-whitehouse.archives.gov/news/releases/2005/06/20050629-12.html

White House, President George W. Bush, *Open Letter on the President's Position on Climate Change*, February 7, 2007; http://georgewbush-whitehouse.archives.gov/news/releases/2007/02/20070207-5.html

White House, President George W. Bush, *President Bush Discusses Global Climate Change*, June 11, 2001; http://georgewbush-whitehouse.archives.gov/news/releases/2001/06/20010611-2.html

White House, President George W. Bush, *President Bush Discusses the Budget*, October 15, 2007; http://georgewbush-whitehouse.archives.gov/news/releases/2007/10/20071015-5.html

White House, President George W, Bush, President Bush Participates in Major Economies Meeting on Energy Security and Climate Change, September 28, 2007; http://georgewbush-whitehouse.archives.gov/news/releases/2007/09/20070928-2.html

White House, President George W. Bush, *Press Conference by the President*, March 29, 2001; http://georgewbush-whitehouse.archives.gov/news/releases/2001/03/20010329.html

White House, President George W. Bush, *Text of a Letter from the President to Senators Hagel, Helms, Craig and Roberts*, March 13, 2001; http://georgewbush-whitehouse.archives.gov/news/releases/2001/03/20010314.html

White House, *Restoring the Quality of Our Environment*, p. 112, November, 1965; http://dge.stanford.edu/labs/caldeiralab/Caldeira%20downloads/PSAC,%201965,%20Restoring%20the%20Quality%20of%20Our%20Environment.pdf

William J. Clinton Presidential Library and Museum, *Vice-President's Remarks on U.S. Signing of the Kyoto Protoco*, November 12, 1998; http://clinton5.nara.gov/CEQ/19981112–7936.html

Woodwell, G.M., MacDonald, G.J., Revelle, R., and Keeling, C.D., The Carbon Dioxide Problem: Implications for Policy in the Management of Energy and Other Resources, July 1979; http://graphics8.nytimes.com/packages/pdf/science/woodwellreport.pdf

York, G., China Emerges as Rock Star at Durban Climate Summit, *Globe and Mail*, December 7, 2011; http://www.theglobeandmail.com/news/world/china-provides-a-glimmer-of-hope-in-durban/article2261157/; accessed December 2011

Notes

1 Hecht and Tirpak, 1995, p. 373.
2 Mintzer and Leonard, 1994, pp. 33–5.
3 See for example, IEA, Boxer to Focus on Climate Science, *Energy Efficiency and Climate Change News*, December 8–14, 2010.
4 Borione and Ripert, in Mintzer and Leonard, 1994, p. 82; *Earth Negotiations Bulletin*, 12 (21), April 10, 1995, p. 11; for interference by OPEC at later COPs, see *Earth Negotiations Bulletin*, 12 (123), November 8, 1999, p. 15; *Earth Negotiations Bulletin*, 12 (472), June 14, 2010, p. 1; Kjellen, in Mintzer and Leonard, 1994, pp.156–7; Goldemberg, in Mintzer and Leonard, 1994, p.143; Rahman and Roncerel, in Mintzer and Leonard, pp. 262, 268; Cass, 2006, pp. 39–40.
5 *Earth Negotiations Bulletin*, 12 (21), April 10, 1995; Mintzer and Leonard, pp. 35, 39; Borione and Ripert, in Mintzer and Leonard, p. 83.
6 Dasgupta, in Mintzer and Leonard, 1994, p. 132.
7 US House of Representatives, May 19, 1995, pp. 171–2.
8 *Earth Negotiations Bulletin*, 12 (534), December 13, 2011, p. 4.
9 Testimony of Roger Revelle, US Congress, House 84 H1526-5, Committee on Appropriations, Hearings on Second Supplemental Appropriation Bill (1956), p. 467; cf. Weart 2008.
10 Bodansky, in Mintzer and Leonard, 1994, p. 46; the proceedings were reported in *Implications of Rising Carbon Dioxide Content of the Atmosphere*.
11 White House, November 1965, pp. 112–13, 126–7.
12 Schneider, 2009, p. 84.
13 National Research Council, National Academy of Sciences, July 1979.
14 Woodwell *et al.*, 1979, pp. 7–8, 11; also see Hecht and Tirpak, 1995, and Bolin, 2007, pp. 33–4; these articles summarize the early initiatives that led to political action on climate change.
15 Davidson, November 20, 2012.
16 Murse (blog).
17 Davidson, November 20, 2012.
18 He is currently Director for Sustainable Development, Office of Research and Development, at the EPA.
19 He is currently a Senior Fellow of the World Resources Institute and an Associate of the International Institute of Sustainable Development in Canada.
20 Hecht and Tirpak, 1995, pp. 380–1, 400.
21 Pearce, 2005, p. 52.
22 Hecht and Tirpak, 1995, p. 401.
23 Cass, 2006, p. 32.
24 Shabecoff, January 31, 1989.
25 IPCC, *Report of the Fifth Session of WMO/UNEP Intergovernmental Panel on Climate Change*, p. 5.
26 Schneider, 2009, pp. 86–7.
27 Hecht and Tirpak, 1995, p. 381.
28 Shabecoff, June 24, 1988; Hansen's use of the term "global warming" at the Senate hearing popularized its use.
29 *New York Times*, April 21, 1990; also see Bodansky, in Mintzer and Leonard, 1994, p. 49; Hecht and Tirpak, 1995, p. 383.

30 Hecht and Tirpak, 1995, p. 393.
31 Shabecoff, May 13, 1989.
32 Hecht and Tirpak, 1995, p. 394; IPCC, *Report of the Third Session of WMO/UNEP Intergovernmental Panel on Climate Change*, p. 3.
33 Shabecoff, May 13, 1989.
34 Gelb, February 10, 1991.
35 Nitze, in Mintzer and Leonard, 1994, p. 193; Shabecoff, April 17, 1990.
36 Nitze, in Mintzer and Leonard, 1994, pp. 192–4; Sebenius, in Mintzer and Leonard, 1994, p. 288; Gelbe, February 10, 1991.
37 Hecht and Tirpak, 1995, p. 402.
38 Nitze, in Mintzer and Leonard, 1994, p. 192.
39 Bodansky, in Mintzer and Leonard, 1994, pp. 55–6; Borione and Ripert, in Mintzer and Leonard, 1994, p. 80; Kjellen, in Mintzer and Leonard, 1994, pp. 151, 158, 164; Tariq, in Mintzer and Leonard, 1994, p. 209; Rahman and Roncerel, in Mintzer and Leonard, 1994, p. 268; Cass, 2006, p. 69.
40 Rahman and Roncerel, in Mintzer and Leonard, 1994, p. 254.
41 IPCC, *Report of the Third Session of WMO/UNEP Intergovernmental Panel on Climate Change*, p. 3.
42 Ibid., p. 4.
43 *New York Times*, April 21, 1990.
44 Shabecoff, April 18, 1990.
45 Bolin, 2007, p. 60.
46 United Nations, *Intergovernmental Negotiating Committee for a Framework Convention on Climate Change, First Session*, p. 4; Cass, 2006, p. 78.
47 United Nations, *Intergovernmental Negotiating Committee for a Framework Convention on Climate Change, Second Session, Item 2 of the Provisional Agenda*, p. 94.
48 Dasgupta, in Mintzer and Leonard, 1994, p. 134.
49 Kjellen, in Mintzer and Leonard, 1994, p. 162; also see p. 160.
50 Bodansky, in Mintzer and Leonard, 1994, pp. 61, 63, 65, 68; Dasgupta, in Mintzer and Leonard, 1994, pp. 136–7, 141; Goldemberg, in Mintzer and Leonard, 1994, pp. 179, 181–2; Hyder, in Mintzer and Leonard, 1994, p. 209. There was some support for the US position from the UK and Japan.
51 Sebenius, in Mintzer and Leonard, 1994, p. 281.
52 Goldemberg, in Mintzer and Leonard, 1994, p. 182.
53 Ibid., p. 179.
54 Hecht and Tirpak, 1995, p. 391.
55 Schneider, 2009, pp. 128–30.
56 Borione and Ripert, in Mintzer and Leonard, 1994, p. 91.
57 Hecht and Tirpak, 1995, pp. 371–2.
58 Ibid., p. 392.
59 Schneider, 2009, p. 132.
60 Dasgupta, in Mintzer and Leonard, 1994, p. 143.
61 This result is surprising, as the EU would have been expected to be the first. However, the EU was having its own problems. Germany and other nations were refusing to ratify if the EU did not have a common policy to deal with climate change. A European policy was not reached in time, but Germany ceded, and the EU ratified at the end of the year; see Cass, 2006, p. 113.
62 See Cass, 2006, pp. 95, 120; his last name is sometimes incorrectly reported as "Pomerantz."
63 Ibid., p. 97.
64 *The Climate Change Action Plan.*
65 Cass, 2006, p. 98.
66 Ibid., p. 96.
67 US House of Representatives, March 4, 1998, p. 1.
68 Ibid., see p. 61 for example.
69 US House of Representatives, March 21, 1995, p. 1.
70 Ibid., p. 10.

71 Ibid., p. 61.
72 Ibid., p. 136.
73 Ibid., p. 74; also see p. 38.
74 *Earth Negotiations Bulletin*, 12 (21), April 10, 1995, p. 11.
75 US House of Representatives, May 19, 1995, pp. 116–17; also see pp. 120–1, 174.
76 Ibid., p. 173.
77 Ibid., p. 112; for other statements also see, pp. 113, 127, 131–2, 139, 142.
78 Ibid., pp. 135–42, 170–8.
79 Ibid., pp. 177–8.
80 US Senate, *Byrd–Hagel Resolution*.
81 Cass, 2006, p. 129.
82 See US House of Representatives, 1998, p. 2. Tim Wirth went to work with the new UN Foundation that had been established by a grant from Ted Turner. A quirky reference to Ted Turner was made at the Subcommittee on Energy and Power, where Dan Schaefer referred to Turner as "Henry Fonda's daughter's husband."
83 US Department of State, September 4, 1997; also see Bolin, 2007, p. 162.
84 *Earth Negotiations Bulletin*, 12 (55), August 11, 1997, p. 13.
85 *Earth Negotiations Bulletin*, 12 (76), December 13, 1997, p. 3.
86 Ibid., pp. 5–6.
87 Ibid., p. 13.
88 IPCC, *First Assessment Report, Working Group III: The IPCC Response Strategies*, p. 244, see Section 9.5.3, pp. 240–1; the mechanism had been vaguely referred to in the UNFCCC signed at Rio (Article 4.2a).
89 George Bush, April 1990; also see IPCC, *Report of the Third Session of WMO/UNEP Intergovernmental Panel on Climate Change*, p. 3.
90 *The Climate Change Action Plan*.
91 US House of Representatives, March 21, 1995, p. 28; also see p. 20.
92 US House of Representatives, May 19, 1995, p. 117.
93 Ibid., pp. 172, 176.
94 Ibid., p. 142.
95 See Cass, 2006, p. 121.
96 Convery *et al.*, March 2008, pp. 7–8; see UNFCCC, October 29, 1996, p. 48.
97 European Commission, October 1991; also see Usui, 2005, p. 16; Sebenius, in Mintzer and Leonard, 1994, p. 294; Cass, 2006, pp. 112–13, 146–8; US Department of State, OES, September 5, 1997, paragraph 31.
98 Cass, 2006, p. 70.
99 Mackenzie, D., 2008, p. 5; Cole, 2009, p. 3.
100 Harrison, November 2007, p. 102.
101 *Earth Negotiations Bulletin*, 12 (55), August 11, 1997, p. 12.
102 US Department of State, September 15, 1997.
103 US Department of State, OES, September 5, 1997, paragraph 12; for an overview of EU policy see paragraphs 41–3; this document was approved by Wirth.
104 *Earth Negotiations Bulletin*, 12 (38), July 22, 1996, p. 13; also see Grubb, M., Hourcade, J.-C., and Oberthur, S., *Keeping Kyoto, Climate Strategies* July 2001, p. 5.
105 Warrick, June 28, 1997.
106 *Earth Negotiations Bulletin*, 12 (76), December 13, 1997, p. 2.
107 US Department of State, September 4, 1997.
108 US Department of State, October 2, 1997.
109 Weekly Compilation of Presidential Documents, October 27, 1997, p. 1631.
110 Cass, 2006, p. 159.
111 Bolin, 2007, p. 153.
112 Cass, 2006, p. 144.
113 See discussions below, and US House of Representatives, March 4, 1998, p. 90.
114 Ibid., p. 16.
115 *Earth Negotiations Bulletin*, 12 (76), December 13, 1997, p. 15.

116 Cass, 2006, p. 165.
117 US House of Representatives, May 19, 1995, p. 176.
118 Ibid., p. 129; also see pp. 171, 176.
119 Ibid., p. 175.
120 Ibid., p. 117.
121 US House of Representatives, March 4, 1998, p. 20.
122 Ibid., p. 15.
123 Ibid.; also see p. 20.
124 Eizenstat, October 28, 1998.
125 US House of Representatives, 1998, p. 16.
126 Ibid.
127 Ibid., p. 47.
128 Ibid., p. 3.
129 Ibid., p. 69.
130 US Department of State, OES, September 5, 1997, paragraph 23.2; also see US Department of State, September 15, 1997; US Department of State, *The Kyoto Meeting*; US House of Representatives, March 4, 1998, p. 41.
131 For example, see US House of Representatives, March 4, 1998.
132 Eizenstat, October 28, 1998 – the title of his speech was "Fighting Global Warming From Kyoto to Buenos Aires and Beyond."
133 William J. Clinton Presidential Library and Museum, November 12, 1998.
134 For example, see *Earth Negotiations Bulletin*, 12 (209), November 4, 2002, pp. 12, 14.
135 For example, see *Earth Negotiations Bulletin* 12 (189), November 12, 2001, p. 16.
136 Historycentral.com, 2000.
137 White House, President George W. Bush, March 13, 2001.
138 Jehl, April 4, 2001.
139 Criticism of the Bush Administration policy on Kyoto and climate change are presented by Harrison, November 2007, pp. 104–5, 111; Mueller, *et al.* 2001 – the paper gives the pretense that it presents alternatives to the Kyoto Protocol but, in fact, it only condemns the US position; an interesting article on ethical, legal and fairness issues in the US position is presented by Posner and Sunstein, August 2007; for statements by President Bush see for example, White House, President George W. Bush, March 29, 2001; White House, President George W. Bush, June 29, 2005; White House, President George W. Bush, October 15, 2007; for positive comments on climate change by President Bush see White House, President George W. Bush, June 11, 2001; White House, President George W, Bush, *Open Letter on the President's Position on Climate Change*, February 7, 2007; and White House, President George W, Bush, September 28, 2007.
140 McCright and Dunlap, 2010, p. 101, and references therein; also Revkin, October 19, 2004.
141 Mackenzie, D., 2002; criticism of the Bush Administration on climate change is well documented in McCright and Dunlap, 2010, pp. 101–2, 115–21, 125, and references therein.
142 Luntz, p. 131.
143 Luntz, p. 137.
144 IPCC, *Report of the Nineteenth Session of the Intergovernmental Panel on Climate Change*, Appendix A.
145 Bolin, 2002, p. 1235.
146 Bolin, 2007, pp. 185–6.
147 See, for example, Bolin, 2007, p. 186.
148 Lawler, 2002, p. 232.
149 Ibid.

150 Randol, February 1, 2001; also see Gore, April 21, 2002; Mackenzie, 2002; Lean, January 22, 2010; Vidal, September 3, 2010.
151 See: http://www.whitehouse.gov/administration/eop/ceq/
152 Vidal, June 8, 2005.
153 InterAcademy Council, 2010, p. 19; also see p. xiv; *Earth Negotiations Bulletin*, 12 (486), pp. 6, 15, October 18, 2010.
154 UNFCCC, December 18, 2009, p. 2; see also UNFCCC, March 30, 2010, paragraph 5, p. 6.
155 For a list of the original accepting countries, see ibid., p. 5.
156 White House, President Barack Obama, December 18, 2009.
157 *Earth Negotiations Bulletin*, 12 (459), December 22, 2009, pp. 8, 26, 28.
158 For a positive review of Copenhagen see Bodansky, February 12, 2010.
159 *Earth Negotiations Bulletin*, 12 (498), December 13, 2010, p. 28.
160 Blackwell, January 22, 2010.
161 Max, December 5, 2011.
162 UNFCCC, *Principles*, 1992, 3.1, p. 3.
163 Ibid., 3.2, p. 3; Article 10, p. 9.
164 China Climate Change Info-net, December 2011.
165 US House of Representatives, March 4, 1998, pp. 3–4.
166 Rio+20, United States of America, November 1, 2011.
167 Rio+20, European Union and its Member States, November 1, 2011.
168 Rio+20, Group of G77 and China, November 1, 2011.
169 US House of Representatives, March 4, 1998, p. 21; also see US Department of State, OES, September 5, 1997, paragraph 22; Sebenius, in Mintzer and Leonard, 1994, p. 289.
170 All CO_2 levels are given as CO_{2e}, not as C.
171 Kjellen, in Mintzer and Leonard, 1994, p. 171; United Nations, *Intergovernmental Negotiating Committee for a Framework Convention on Climate Change, Third Session, Compilation of Proposals Related to Commitments*, August 15, 1991 pp. 10, 13, 19, 25, 34; *Earth Negotiations Bulletin*, 12 (76), December 13, 1997, p. 13; *Earth Negotiations Bulletin*, 12 (395), December 15, 2008, p. 13.
172 Nitze, in Mintzer and Leonard, 1994, p. 196.
173 Ibid.
174 For a recent discussion, see Helm, November 8, 2012.
175 *China Daily*, July 22, 2010.
176 Bolin, 2007, p. 108.
177 Stevens, February 21, 1995.
178 Some experts and nations have made the argument of "carbon debt" or "carbon justice" regarding who should pay for the consequences of climate change (see Mueller 2001, *et al.*, p. 4). But the intent of the fight against climate change should be to decarbonize, not demonize.
179 Goldemberg, in Mintzer and Leonard, 1994, p. 182.
180 Nitze, in Mintzer and Leonard, 1994, p. 195.
181 Broder, December 11, 2011.
182 The government has even established a website in English to summarize the latest domestic and international developments on climate change – China Climate Change Info-net.
183 Hu Jintao, September 2009, p. 8; also see China Climate Change Info-net.
184 Feinberg, January 11, 2011.
185 As indicated by clean technology patents; see Ren21, *Global Status Report*, 2010, p. 52.
186 Ren21, 2007, pp. 40, 41.
187 Ren21, 2010, p. 10; also see Ren21 *Global Status Report*, 2011, p. 73.
188 Ren21, 2011, p. 11.

189 Buckley, *PlanetArk*, February 28, 2011.
190 The exportation of manufacturing facilities to countries without carbon regulations is called "carbon leakage." For a review see Grubb, *et al.*, August 2009, Section 5, pp. 24ff; Grubb *et al.*, Section 4, pp. 40ff; Helm, November 11, 2012; this is a real problem, especially for energy-intensive industries, such as aluminum, concrete, chloralkali and steel.
191 The Persian Gulf countries are an even more extreme example of this with their oil exports.
192 See Helm, 2008.
193 Kjellen, in Mintzer and Leonard, 1994, p. 164.
194 *Earth Negotiations Bulletin*, 12 (97), November 16, 1998, p. 3.
195 Lou and Lawrence, June 4, 2007.
196 Revkin, September 2010.
197 US Department of State, OES, September 5, 1997, paragraph 24; also see US House of Representatives, March 21, 1995, p. 95.
198 *Earth Negotiations Bulletin*, 12 (260), December 20, 2004, p. 16.
199 *Earth Negotiations Bulletin*, 12 (21), May 19, 2005, p. 1.
200 *Earth Negotiations Bulletin*, 12 (291), December 12, 2005, p. 17.
201 *Earth Negotiations Bulletin*, 12 (318), November 20, 2006, p. 11.
202 *Earth Negotiations Bulletin*, 12 (354), December 18, 2007, pp. 16, 20; generally, the Americans were viewed as obstructing progress at Bali (see Pooley, 2010, pp. 3, 14–15).
203 *Earth Negotiations Bulletin*, 12 (395), December 15, 2008, p. 13.
204 Harvey, October 29, 2010.
205 *Earth Negotiations Bulletin*, 12 (498), December 13, 2010, p. 29.
206 Stanway, September 13, 2010.
207 Ministry of Foreign Affairs of the People's Republic of China, November 10, 2010.
208 York, December 7, 2011.
209 *Earth Negotiations Bulletin*, 12 (534), December 13, 2011, p. 27.
210 Van Renssen, December 15, 2011.
211 *Earth Negotiations Bulletin*, 12 (534), December 13, 2011, p. 31.
212 Clinton, December 14, 2011.
213 Lan and Jing, December 12, 2011.
214 *Earth Negotiations Bulletin*, 12 (546), May 28, 2012, p. 24.
215 *Earth Negotiations Bulletin*, 12 (567), December 11, 2012, p. 28; also see *Earth Negotiations Bulletin*, 12 (546), May 28, 2012, pp. 24–5; *Guardian*, November 22, 2012.
216 Buckley, November 23, 2010.
217 For data see BP, 2011: China coal production = 1,553 million tonnes of equivalent oil; OPEC oil production = 1,575 million tonnes of oil. For an overview of the Chinese coal industry see Aden *et al.*, 2009; China Electricity Council, *News*; Lim, L., China's Coal Fueled-Boom has Costs, NPR News, May 2, 2007.
218 BP, 2011.
219 For a detailed review of the messy politics of climate change in Canada see Simpson *et al.*, 2007.
220 For example, the interview by the Premier, Ralph Klein, on October 23, 2002; http://www.ctv.ca/servlet/ArticleNews/story/CTVNews/1035395745106_43/; accessed November 2011.
221 Mason, December 16, 2009.
222 Government of Alberta, Environment, Legislation/Guidelines; http://environment.alberta.ca/03147.html.
223 CBC News, November 28, 2007.
224 Ontario, September 3, 2009.
225 Ontario, January 9, 2013; also see *Renewables International*, January 11, 2013.
226 CBC News, December 12, 2011; De Souza, December 14, 2012.
227 All web-based references were confirmed (accessed) in June 2013, unless otherwise stated.

4 Unsportsmanlike conduct

EU climate change policy

The most valuable player in the carbon game has been the European Union, which has led the world in stats on reducing GHG emissions. However, its exceptional performance is tainted as its reductions had little to do with climate change. In this chapter we will uncover how the EU countries, playing in a league of their own, won the carbon game by subsidizing their industries, thereby enhancing their own competitive position. Among these policy "steroids" are the feed-in-tariff (FIT) and Emissions Trading System (ETS).

A FIT over solar subsidies

A feed-in-tariff (FIT) encourages renewable power projects by providing long-term contracts at a premium over standard rates over a fixed period of time, usually in the range of ten to 20 years. There are two basic FIT instruments which we will refer to as technology-neutral and technology-specific; the intent of the latter is to balance the economic disparity between the various renewable options. Among the major renewable energy sources, solar energy is the least economic, so it gets the largest subsidies under a technology-specific FIT. An excellent review of the FIT program was prepared by NREL (National Renewable Energy Laboratory) of the US Department of Energy. The request for the study, interestingly enough, came from the Solar Energy Technologies Program of the US Department of Energy. The experts at NREL presented an overview of all the different types of FIT programs, including the popular technology-specific FIT. With regard to the latter, the authors reported that its purpose was to encourage diversity of renewable energy sources and that it can "increase the overall costs,"[1] but if the goal is jobs and economic development, technology-specific FIT programs can be used.[2] NREL, however, never stated that the technology-specific FIT is a preferred option to reduce emissions.

The FIT program was first introduced in the United States in 1978. When the government passed the Public Utilities Regulatory Policy Act to promote renewable energy, states such as California issued Standard Offer Contracts for renewable energy resulting in 1.2 GW of wind energy being installed by the middle 1980s. It was Germany, though, that made the FIT program famous. In

Table 4.1. Feed-in-tariff incentives for renewable fuels in Germany, Spain and Ontario.

	Germany	Spain	Ontario
Biomass	€77/MWh	€81/MWh	$130/MWh
Wind	€60/MWh	€61/MWh	$135/MWh
Solar	€284/MWh	€269/MWh	$443/MWh

Source: US Deptartment of Energy, 2010, p. 137; for Ontario also see OPA, *Renewable Energy Feed-in-Tariff Program*, *FIT Price Schedule*; http://fit.powerauthority.on.ca/program-archives; also see C.D. Howe Institute Commentary, February 2009, pp. 17–18.

1991 the *Stromeinspeisungsgesetz*[3] provided a technology-neutral FIT, which also spurred on early wind energy investment. However, in 2000, Germany adjusted its FIT program to create the first technology-specific subsidies for electricity generation.[4] With the introduction of the new FIT program, the solar subsidies soared from $0.08/kWh to almost $0.50/kWh.[5]

Such programs are now especially popular across the EU-27, and have been adopted by 21 members.[6] The renewable energy source (RES) investments have been a boon to the European economy – over half a million jobs were created in an industry sector that had not existed before; in 2009, the renewable industry sector had annual sales of €70 billion.[7] Through such policies, the RES sector in Germany, with over 370,000 jobs, is becoming the largest manufacturing employer in the nation.[8] As part of its strategy, the Ontario government had also introduced a FIT program, similar to Germany's (Table 4.1).

The widely acclaimed FIT program[9] has been declared "successful." But what is the definition of success? "Success," apparently, means implementation of more solar power in these regions, as such FIT programs have accounted for 75 percent of all solar photovoltaic (PV) installations in the world;[10] and sunny Germany has far exceeded all other countries in the adoption of solar.[11] In an ironic twist of fate, the FIT has transformed solar into an unsustainable industry, for only as long as the artificial FIT subsidies exist from fickle governments can the industry survive;[12] for example, on May 30, 2013, Ontario cancelled the FIT program for large projects.[13]

At first glance, the technology-specific FIT appears to be an effective policy in fighting climate change, but there is a serious flaw. This apparent "success" has come at a high cost as every taxpayer has been forced to support the solar industry. The fiscal result is the most expensive emissions reduction programs in the world. We will use Ontario as an example. If we convert the difference in the subsidies of the Ontario FIT program between solar ($443/MWh) and wind ($135/MWh) or biomass ($130/MWh), it translates into a carbon value of $354 per metric tonne of CO_2.[14] This huge subsidy for solar[15] is just the extra cost of solar over other renewable energies! The disproportionate subsidy of the technology-specific FIT limits the installation of more economic technologies; for example, a wind or biomass project in Ontario which could commercially succeed with a $0.20/kWh rate will not proceed, while a solar project requiring a $0.40/kWh rate *will*, under this incentive system. The C.D. Howe Institute concluded that the Ontario government is "over-investing taxpayers' money in high-cost mitigation technologies and under-investing in low-cost mitigation technologies."[16] To ensure that everyone understands, the problem with this FIT program is not the subsidy over fossil fuels, but the subsidy for solar over other renewable energy sources. The end result is that taxpayers pay several times the carbon price for solar power but gain no further carbon reductions compared to other renewable energy sources. If climate change (rather than support for a particular industry) is the priority, government incentives for carbon-neutral projects must be technology-neutral as well.

While the solar industry has certainly flourished with its FIT subsidies, overall implementation of renewable energy sources (RES) within the EU has not been so successful. Back in 1997, a White Paper[17] had proposed a renewable target of 12 percent by 2010 (from 5 percent in 1995) for the EU-15, and the target was confirmed by members in October 2002. But several challenges hampered the rollout of the 1997 program:[18]

- the high cost of renewable energy owing to the investment required and the fact that externalities (the "external" cost of the different energy sources, particularly their long-term impact on health or the environment) have not been taken into account, which gives fossil fuels an artificial advantage;
- administrative problems resulting from installation procedures and the decentralized nature of most renewable energy applications;
- the opaque and/or discriminatory rules governing grid access;
- inadequate information for suppliers, customers and installers;
- the fact that the 12 percent target is expressed as a percentage of primary energy, which puts wind power at a disadvantage (a sector that has experienced considerable growth during the period in question).

A progress report was issued on May 26, 2004, and the conclusion was that renewable penetration would only be 10 percent by 2010.[19] The overall lack of progress led the European Council in March 2006 to ask for another plan to promote RES. The Renewable Energy Roadmap, issued on January 10, 2007,[20] bluntly stated the reason for the failure of this critical program for the EU was policy.

Nevertheless, Europe is the self-proclaimed leader of RES technologies:[21]

> Europe is the pioneer in developing and implementing modern renewable energy techniques. Western Europe, with its 16% of world energy consumption, accounted for 31% of the world increase in electricity generation from biomass between 1990 and 2000; 48% of the increase in small hydropower; and 79% of the increase in wind power. The European Union and its Member States have pioneered the policy and regulatory arrangements, such as targets, and the financial schemes needed to drive renewable energy forward. European companies lead the world in renewable energy technology.

However, the facts do not support such claims. Their policies have not encouraged renewable electrical investments more than countries without Kyoto targets or FIT programs; for example, the RES electrical capacity of China was the same as

Table 4.2. Renewable energy attractiveness statistics for the top ten nations (relative ranking): 2013.

	Overall	Wind	Solar PV	Biomass	Geothermal
China	70	76	65	59	50
Germany	66	68	61	68	58
US	65	63	70	61	67
India	62	61	65	59	43
France	57	59	55	58	35
UK	55	62	42	57	35
Japan	53	51	61	43	49
Canada	53	62	40	50	35
Italy	52	53	53	49	57
Australia	51	50	54	44	58

Source: Ernst & young, February 2013, p. 16.
Note: The index is weighted 35 percent from an infrastructure index and 65 percent from a technology index; the latter includes: power agreement structure and price, taxes, grants, growth potential, current capacity, resource availability and project size (see p. 36 for details); for other country ratings on climate change see RICS Research, September 2011; Yale Center for Environmental Law & Policy, 2012; Climate Institute, *G20 Low Carbon Competitiveness* February 2013; for G8 ratings see National Round Table on the Environment and the Economy, May 20, 2010, p. 15; and ECOFYS, *G8 Climate Scorecard.*

the entire EU-27 in 2010. In an annual survey of the legislation, natural resources, and other factors that encourage renewable energy investment, only four of the top ten nations are European (Table 4.2).

The emissions trading "subsidy"

A pioneering aspect of the EU climate change policy was the creation of the carbon market through the Emissions Trading System (ETS).[22] Previously, I discussed the importance of the joint implementation and the flexible mechanisms to the US, which paid a heavy price to have them included in the Kyoto Protocol; specifically, the US was saddled with similar reduction targets as the EU, which, under the circumstances, gave the EU a major competitive advantage. Adding insult to injury, the Europeans then turned the flexible mechanisms, so dear to the Americans, into a component of their own competitiveness strategy.

The tools of the Kyoto flexible mechanisms are:

1. *Emissions Trading* – an economic instrument to encourage private sector participation. The intent was to take full advantage of the lowest cost options, and those emitters with higher costs could "buy" the excess reductions from the low emitters. The innovative scheme is defined as follows:[23]

   Parties with commitments under the Kyoto Protocol (Annex B Parties) have accepted targets for limiting or reducing emissions. These targets are expressed as levels of allowed emissions, or "assigned amounts," over the 2008–2012 commitment period. The allowed emissions are divided into "assigned amount units" (AAUs). Emissions trading, as set out in Article 17 of the Kyoto Protocol, allows countries that have emission units to spare – emissions permitted them but not "used" – to sell this excess capacity to countries that are over their targets.

Thus, a new commodity was created in the form of emission reductions or removals. Since CO_2 is the principal greenhouse gas, people speak simply of trading in carbon. Carbon is now tracked and traded like any other commodity. This is known as the "carbon market."

2. *Clean Development Mechanism* (CDM)[24] – another option to lower costs was to allow an Annex I country (industrialized nation that has a Kyoto target) to create an emissions reduction project in a developing country, which would be counted towards meeting the Annex I country Kyoto target ("offsets"[25]). Consequently, a project to reduce emissions could be built by Germany, for example, in South Africa, and any carbon reduction would be counted towards Germany's Kyoto target. Such projects would be issued a certified emissions reduction (CER) unit.[26]
3. *Joint Implementation* (JI) – a similar option to the CDM, but the projects take place in other Annex I countries. By doing so, the project country has "sold" the reduction to the investing country, and the former cannot include such projects in their own national totals. The JI credits are named emissions reduction units (ERUs).
4. *Land Use, Land Use Change & Forestry* (LULUCF) – an incentive to promote reforestation, forest protection, and related projects as "sinks" for CO_2 (i.e., growing biomass removes CO_2 from the atmosphere). To add yet another acronym to the carbon currency portfolio, such projects are issued a removal unit (RMU). A related program is Reducing Emissions from Deforestation and Degradation (REDD) but it only applies to developing countries.

In Europe, these mechanisms came together under the Emissions Trading System (ETS):[27]

> The aim of the EU Emissions Trading System (EU ETS) is to help EU Member States achieve their commitments to limit or reduce greenhouse gas emissions in a cost-effective way. Allowing participating companies to buy or sell emission allowances means that emission cuts can be achieved at least cost. The EU ETS is the cornerstone of the EU's strategy for fighting climate change ... At the end of each year installations must surrender allowances equivalent to their emissions. Companies that keep their emissions below the level of their allowances can sell their excess allowances. Those facing difficulty in keeping their emissions in line with their allowances have a choice between taking measures to reduce their own emissions – such as investing in more efficient technology or using less carbon-intensive energy sources – or buying the extra allowances they need on the market, or a combination of the two. Such choices are likely to be determined by relative costs. In this way, emissions are reduced wherever it is most cost-effective to do so.

Table 4.3. Phase I of the ETS: ratio of CO_2 emissions for selected nations: 2007/2005.

	Total	ETS
Germany	-2.7%	+2.5%
UK	-2.7%	+5.8%
France	-3.7%	-3.5%
Italy	-2.1%	+0.2%
Spain	+0.5%	+1.6%
EU	-1.4%	+1.9%
US	+0.5%	n.a.

Source: Total data from BP; ETS data from EEA Report, No. 13/2008, p. 15.

The ETS is based on the Crocker–Dales cap-and-trade system, and focuses on major "smoke-stack" sources of CO_2 emissions; these companies accounted for 43 percent of total EU emissions in 2007.[28] The remaining emissions came mainly from smaller industries and tailpipe (transportation) emissions. The popular market strategy sets an upper annual target of GHG emissions for a country on a declining basis, which is the "cap." The national cap is then subdivided on a facility level. Carbon credits (or emissions allowances) are then issued to companies based on the allowable emissions. Companies must have credits at least equal to their emissions each year. If a business has excess credits because it has reduced its emissions more than required to meet the cap, then these can be traded.[29] Companies can buy these extra carbon credits to allow them to keep on emitting without reducing emissions. Overall, emissions should have been reduced. These have been granted over three periods of National Allocation Plans (NAP): Phase I lasted until the end of 2007; Phase II to 2012; and Phase III to 2020.[30] The segmented scheme allowed for corrections to be made as required.[31]

Phase I: 2005–7

With Phase I completed (2007), the impact of the program on climate change can be assessed[32] (Table 4.3; data from the US are included for comparison purposes).

During the short three-year time span that the ETS was active, the EU did very well in terms of total emissions (column 1 – total all emissions, including companies within the ETS). However, the ETS was not a contributing factor in this overall performance as industrial emissions from major companies actually increased in most countries (column 2 – only includes companies within the ETS). Cap-and-trade systems are supposed to guarantee a specific level of emissions reductions. What happened?

The results show that the trading system of the ETS had been converted into a trading subsidy. This is accomplished by issuing more credits than there were emissions. Initial NAP submissions by the member states were so flagrantly high that even the EU had to step in to reduce the emissions allowances submitted. While the EU forced major reductions in the NAP, the approved allocations were still in excess of historic emissions.[33] The carbon credits allocated for 2005 were over the emissions by 4 percent. While the allowances or credits declined over the three years, in 2007, they were still over 2005 actual emissions levels. Consequently, there was no incentive for the major facilities to reduce their emissions. The failure of the ETS was largely caused by the EU-12 (mostly the countries of Eastern Europe), where the over-issuing of credits was 14 percent higher than actual emissions. But the EU-15 members (mostly the countries of Western Europe) were also part of the conspiracy. The worst offenders among the EU-15 were France by 17 percent and Sweden by 24 percent. Even the leader in emissions reductions, Germany, issued an average of 3 percent excess credits over the three-year period. On the positive side were the UK, whose credits were 16 percent below emissions, Spain 9 percent lower, and Italy 8 percent lower.[34] The generous subsidies, which are what the credits turned out to be, were a disincentive for industry to reduce emissions in most EU countries.

The failure of Phase I has been attributed to the trial nature of the ETS, as this was the first major regulated carbon trading scheme in the world, and the actual emissions by individual facilities were largely unknown. However, the UK had opened a pilot system in 2002 and had experienced the same excessive allocation of credits.[35] The EU should have been on the lookout for such problems. In any case, the intent of Phase I was never to be a simple trial, according to the European Commission: "There is a broad consensus that the EU Emissions Trading Scheme (ETS) will have a positive effect of the renewable energy uptake in the EU from 2005 onwards."[36]

The gross, arguably negligent, government subsidies destroyed the intention of the entire program.[37] Some industries made money off the credits while their emissions continued unabated. A joint IEA/OECD study criticized the subsidies given especially to the electricity industry ("windfall" profits);[38] another analysis concluded the electricity companies to "have profited by many billions of euros" through Phase I[39] and, in the *Report on International Competitiveness* carried out for the European Commission, these "windfall profits" were also highlighted.[40]

Many readers are likely puzzled on how these generous allocations, which the utilities received at no cost, could have resulted in higher electricity rates. In twisted capitalistic logic, the carbon markets created a "value" for carbon which was an "opportunity" cost to the company. While the economic pretext of the carbon value of the allowances has some theoretical validity, from a practical point of view the utilities exploited the situation. At the end of the day, the ETS only provided electricity rate hikes to European consumers.

While the electricity sector gained huge profits from the European climate change program (and did nothing to reduce emissions), most other industries also benefited from the ETS.[41] The ETS was simply an economic subsidy disguised as an environmental initiative. Characterizing Phase 1 as a total disaster, however, may be too harsh.[42] The formation of the carbon market is an accomplishment in its own right. At least carbon emissions now had a value associated with them.

The carbon value is established on a commodity exchange, much like currency trading, where the allowances are the carbon currency. The "eco-euros," more properly called the Kyoto Units – the EUA ("EU Allowance"; equivalent to the AAU), CER unit, ERU and RMU (*vide supra*) – are all equivalent to the reduction of one Mg^{43} of CO_{2e}. The EUA is the principle trading unit, followed by the CER unit. Their value is market driven; in this case, by the perceived needs of industry to utilize them to meet their emissions targets. While only companies actually require them, market traders and speculators often drive their value. The carbon price is influenced by economic activity, fuel pricing and even the weather.[44] The EU also influences the value of these "eco-euros" through issuance of the number of EUA to member states.

In early Phase I, the EUA began to trade at €7 and reached a price of €30, but when the oversupply of credits was announced in June 2006, the price collapsed and carbon became worthless.[45] Since the EUA had been issued gratis, no industry lost money. The big loser was the environment as there was no incentive to reduce GHG emissions.[46] The EUA for the second trading period (Phase II) rebounded, peaking at €25 in August 2008. The economic crisis, though, took industrial output down, reducing GHG emissions in the short term, resulting in the collapse of the EUA which fell to €8 in March 2009. Pricing was near €15 for most of 2010, but when economic turmoil hit again near the end of 2011, the value had dropped back to €8 and stayed at this depressed level through most of 2012.[47] Continued economic woes saw the EUA fall to €3 in early 2013, which recovered slightly by the summer. Yet the carbon market has been an *economic* success, soaring from $8 billion in 2005 to $149 billion in 2011.[48]

Over half a trillion dollars have flowed into the carbon market created by the ETS, but where did the money go? The 2012 World Bank report stated: "A considerable portion of the trades is primarily motivated by hedging, portfolio adjustments, profit taking, and arbitrage."[49] Moreover, in late 2011, the Swiss bank UBS issued a status report of the ETS by its Investment Research unit. Their analysis found that the €210 billion investment had "almost zero impact" on reducing carbon emissions; the ETS results were described as "embarrassing."[50]

Phase II: 2008–12

If we treat Phase I as a pre-season game, then Phase II is when the real game began, and it started off poorly. In November 2006 the European Commission was forced to reject again most of the NAP applications for Phase II because the initial proposals ended up having credits 5 percent higher than 2005, and many countries proposed caps that were even higher than Phase I.[51] Most EU nations were making a mockery of the ETS and their legal commitments to the Kyoto Protocol.

The EU had to once more forcibly reduce the National Allocation Plans (NAPs) for most members, especially among the EU-12 (Eastern Europe). Upon doing so, Poland and six other EU-12 nations sued the EU.[52] The credits for the group were eventually tightened up so that they were 10 percent less than emissions overall. Even so, France and many members of the EU-12 stand out again for having

Table 4.4. Phase II of the ETS: allowances/emissions (%) for selected EU nations: 2008.

	Power	Coke ovens	Ore roasting	Steel	Cement	Pulp & paper	All industries
France	+3	+74	+55	+6	+10	+66	+5
Germany	–30	–39		+76	+3	+12	–18
Italy	–7			+21	+8	+8	–4
Spain	–23	+16	–6	+60	+24	+20	–6
Sweden	–55		–6	+80	+11	+60	+4
UK	–29	+17		+14	+26	+14	–19
EU–15	–21	+7	+3	+44	+11	+20	–10
EU–27	–17	+7	+24	+39	+9	+21	–9

Source: EEA Report, No. 9/2009, p.60.
Note: "+" indicates an over-allocation of allowances (i.e., subsidy of excessive free carbon credits).

allocations much greater than emissions. One lesson learned from the ETS is that if individual jurisdictions are allowed to issue their own credits, the system will be abused. A centralized approach must be utilized.[53]

Problems still persisted with the adjusted allocations in Phase II. The crafty members of the EU found a new way to cheat the system by converting the allowances into subsidies for selected industries through a creative carbon game. The ingenious and devious idea was to assign a disproportionately lower cap to one particular industry so that a higher cap could be assigned to other industry sectors. The allocations for the power generators were reduced the most, 18 percent below actual emissions in 2008, officially as a penalty for passing on the opportunity cost of the credits ("windfall" profits).[54] While the penalty to the electricity industry sounded fair under the circumstances, the fact that this industry received fewer allowances conveniently left more allowances to be assigned to the remaining major industry sectors. Even the EU leaders, Germany[55] and the UK, took advantage of this loop hole (Table 4.4). The carbon credits in Phase II were blatantly being utilized as industry-specific subsidies.

The European Environment Agency (EEA) had acknowledged the failure of Phase I from the over allocation of credit, which was supposed to have been rectified in Phase II. The EEA also recognized the added problem of favored industry sectors in Phase II; in other words, countries were protecting certain industries by an over allocation of allowances.[56] And the European Commission had concluded that Phase II of the ETS "threatened fair competition"[57] within the EU itself.

Phase III: 2013–20

Phase III is supposed to operate between 2013 and 2020. The emissions cap will be reduced by 21 percent by the end of that period, compared to 2005 levels.[58] More of the allowances will be auctioned[59] off instead of being issued at no charge, but a 100 percent auction will not be reached until 2027. The exception is most electricity generators, which will have to bid for all of their carbon credits. A worrisome caveat, however, is that an industry likely to relocate its facilities outside the EU because of these costs would receive free carbon credits.[60] Most industries will probably try to take advantage of this loop hole, and it will be interesting to see how the EU handles the approaching bureaucratic nightmare.

Some are now questioning whether the ETS will continue. Reforms to improve the ailing system were rejected in April 2013; an analyst in the sector stated that the failure to pass the reforms would "make EU ETS irrelevant as an emissions reduction tool for many years to come."[61] A commentary, under the title *ETS, RIP?*, concluded:[62]

> The ETS is the only EU-wide environmental instrument. This week's vote sends a signal that Europeans are more concerned about the costs of that flagship policy than about its benefits ... Europe has long claimed to be leading the global debate on climate change. After the ETS vote, that boast rings hollow.

Hot air

The high profile aspects of European climate change policy have been the FIT programs and ETS. While officially being applied to reduce carbon emissions, the practical result is that they are economic subsidies to improve the competitive positions of domestic industries. The "embarrassing" ETS results appear to be out of character for the EU, which is popularly regarded as the champion of the global effort to reduce GHG emissions. While the emissions have declined in many European nations, the reasons may surprise you.

On the tenth anniversary of the Earth Summit in Rio, May 31, 2002, the EU finally ratified the Kyoto Protocol. The adoption of the Protocol began with the traditional EU-15 (mainly Western European) countries, which had Kyoto commitments as a group and also as individual member states. There is no group target with the larger EU-27 (EU-15 + EU-12), but the later-added EU-12 (mainly Eastern European) countries have individual national targets. A summary of the emissions of the members of the EU-27 is presented in Table 4.5. The original members, the EU-15, accounted for 80 percent of the EU-27 emissions, of which Germany, the UK, Italy, France and Spain make up 60 percent (Tables 4.6, 4.7). Overall, the EU-15 has been able to decrease emissions by 1.4 percent by 2010, compared to 1990[63] – no other industrialized country or region can make such a claim while still showing economic growth. A listing of nations on their progress towards the Kyoto targets illustrates the success of some of the EU nations (Table 4.8).

The EU, also, has apparently kept up its momentum with new targets to reduce emissions. In 2004 new guidelines were proposed with the catchy phrase, "20% by 2020."[64] The proposal was supported in the Renewable Energy Roadmap in 2007, and the new EU "Climate Package" was passed by the parliament on December 17, 2008, which included the following directives for 2020 targets:[65]

Table 4.5. Carbon statistics (annual GHG emissions) by members of the European Environment Agency: 2011 (Tg).

EU-15	3,662	EU-12	940
Germany	917	**Poland**	409
UK	549	**Czech Republic**	141
France	498	**Romania**	124
Italy	494	**Bulgaria**	68
Spain	356	**Hungary**	66
Netherlands	196	**Slovak Republic**	46
Belgium	121	**Lithuania**	21
Greece	119	**Estonia**	21
Austria	82	**Slovenia**	20
Portugal	70	**Latvia**	12
Finland	67	**Cyprus**	9
Sweden	63	**Malta**	4
Ireland	57	**Non-EU**	**539**
Denmark	56	**Turkey**	402
Luxembourg	12	**Norway***	53
		Switzerland	50
		Croatia	29
		Iceland*	5
		Lichtenstein*	0

* Though not officially part of the EU, they are part of the EU-ETS.

Source: EEA Report, No. 6/2012, pp.90–159.
Note: Turkey, Cyprus and Malta do not have any formal GHG reduction targets; also see EEA Report, No. 4/2011, pp. 65, 144 and No. 2/2011, p. vii.

Table 4.6. Carbon statistics (annual CO_2 emissions) for the top five countries in the EU – CO_2 emissions: 1990–2010 (Tg).

	1990	1995	2000	2005	2010
EU	4,481	4,258	4,324	4,493	4,143
Germany	1,031	931	903	883	828
UK	622	588	591	604	548
Italy	435	448	476	501	439
France	412	401	430	431	403
Spain	237	264	336	399	334

Source: BP, 2011.

1. 20 percent reduction in GHGs
2. 20 percent energy efficiency increase
3. 20 percent renewable energy sources (RES).

The new directive was issued on April 23, 2009.[66]

As with most climate change policies within the EU, we must dig a bit deeper to see the real picture. Many would have naturally assumed that the 20 percent should be compared to the original Kyoto target of an 8 percent reduction in emissions. But the original target of 8 percent was only for the EU-15, while the new

Table 4.7. Carbon statistics (annual CO_2 emissions) for the top five countries in the EU – ratios of CO_2 emissions: 2011.

	2011/1990	CO_2 (% of EU)	$teCO_2$ (per capita)	CO_2 (per GDP)
EU	−17%	100%	9.2	392
Germany	−26%	20%	11.2	376
UK	−28%	12%	8.8	293
Italy	−5%	11%	8.1	346
France	−11%	11%	7.6	275
Spain	26%	8%	7.7	372

Source: EEA Report, No. 6/2012, pp.90–159.

Table 4.8. Carbon statistics (annual CO_2 emissions) for selected nations: Kyoto target* and actual.*

Country	Target	Actual
US	−7%	+10%
Canada	−6%	+25%
Germany	−21%	−20%
UK	−13%	−12%
France	+0%	−4%
Spain	+15%	+51%
Italy	−7%	+1%
Poland	−6%	−15%
Russia	0%	−29%
Australia	+8%	+42%

* The average of 2008–12 emissions compared to 1990.
Source: BP, 2013.

target of 20 percent extends over the expanded membership of the EU-27. Only the addition of the new Eastern European members (EU-12) has allowed the EU to set such "ambitious" targets. These "economies in transition" (former Eastern bloc states) of the EU-12 were essentially bribed to join Kyoto by being offered emissions targets that they could not fail to achieve. Their phantom emissions reductions[67] ("hot air") resulted from the economic meltdown in the immediate post-Soviet period just after the Kyoto base year of 1990. The overall result was declining emissions through the 1990s but high emissions on a GDP level.[68] For the average of 2008–12, CO_2 emissions for the major "economies in transition" were 36 percent below 1990 levels[69] and none of these reductions were the result of environmental programs (the largest reductions were recorded by the Ukraine). This "hot air" phenomenon is illustrated by looking at the emissions of Poland[70] and Russia[71] (Russia has greater reductions in emissions since 1990 than Germany) compared to other major countries (Table 4.8). With the addition of the hot air emissions reductions from the EU-12, the 2011 results of the EU-27 are already below 1990 levels by 17.5 percent.[72] So for the EU to achieve its "ambitious" goal, it only has to reduce emissions a few percent more within a decade.

The European Environment Agency (EEA) is already projecting that the EU is "well on track" to achieving the new goal of 20 percent, and the EU has acknowledged that it will reach this target by using "existing policies."[73]

Wall fall & dash-for-gas

Forgetting the hot air emissions of the EU-12 for a moment, how is the original EU-15 doing? Among the members listed in Table 4.8, Germany, the UK and France are not doing badly, and their trends in emissions are impressive since the last quarter of the century (Figure 4.1).

However, there is something odd about the results. The major reductions started around 1980, well before any policies associated with climate change were even considered. Let's look more closely at how Germany and the UK reduced their emissions, with special focus on the period since the Kyoto base year of 1990 (Figure 4.2).

Germany, during this time, had access to its own hot air reductions which were called the "wall fall" reductions.[74] On October 3, 1990, Germany had been reunited. Many of the polluting and non-competitive heavy industries in the East were simply closed over the next few years, accounting for Germany's major reduction in GHG emissions during the first half of the 1990s.[75] Between 1990 and 1994, emissions from the former East Germany had fallen by 43 percent;[76] the reductions more than compensated for rises in emissions in West Germany, where few initiatives had been launched to control them.

Even Germany had second thoughts about taking advantage of the wall fall. It had originally intended for its emissions target to apply to West Germany only. However, the final legislation applied to the united Germany.[77] Furthermore, while no major policies to reduce emissions were introduced by the German government before Kyoto,[78] policies had been passed to provide subsidies to the coal industry (for example, *Kohlepfenning*) which even by European standards were excessive. The coal industry was also supported by the fact that there were in place duties on natural gas, but none on coal.[79] Such coal subsidies continue today[80] in Germany, the fifth largest coal consumer in the world, and the country plans to build ten new coal-fired facilities.[81] The growth of coal in Germany (and other nations in Europe) puts into question the intent of EU environmental policies:[82]

> This coal surge is making a nonsense of EU environmental policies, which politicians like to claim are a model for the rest of the world. European countries had hoped gradually to squeeze dirty coal out of electricity generation. Instead, its market share has been growing.

However, German statistics often show that renewables will soon be the leading source for electricity generation in Germany. While renewables have been growing, coal is double that of renewables; the misleading statistics are derived by separating coal into brown coal (lignite) and hard coal (anthracite).[83]

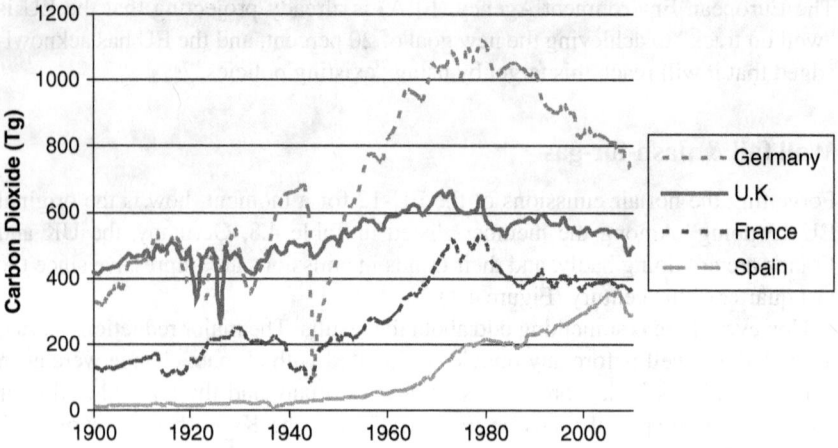

Figure 4.1. Annual CO_2 anthropogenic emissions from major EU emitting countries: 1900–2010.

Source: CDIAC.

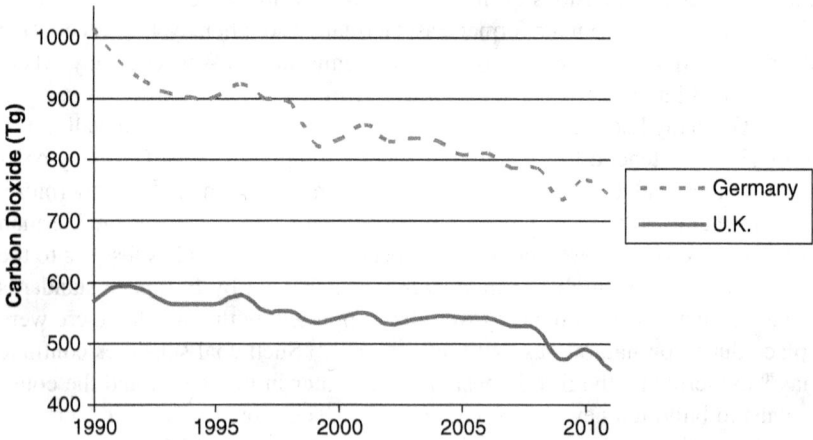

Figure 4.2. Annual CO_2 anthropogenic emissions from Germany and the UK: 1990–2011.

Source: CDIAC.

Germany also sought to block other members of the EU-15 from having access to hot air reductions from Eastern Europe and Russia. On December 11, 1997, Angela Merkel threatened that Germany would not ratify the Kyoto Protocol if emissions trading with the former Soviet states was allowed, and the EU later adopted the motion.[84] Not only did Germany take advantage of its wall fall, it ensured that no one else would have access to similar hot air reductions.

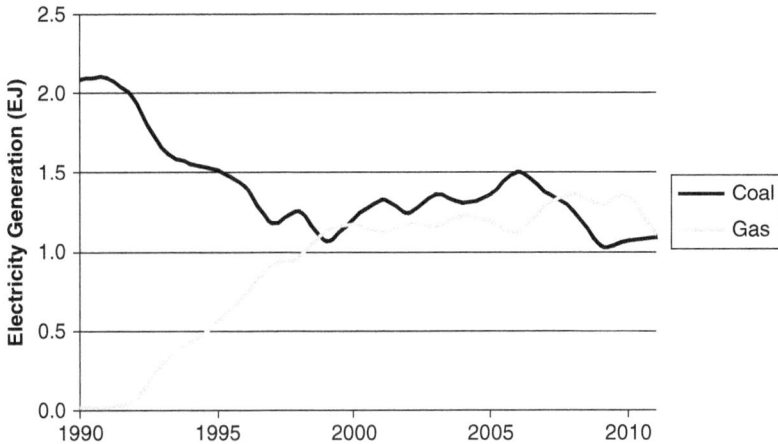

Figure 4.3. Fuel sources for electricity production in the UK: 1990–2011.
Source: UK Department of Energy and Climate Change, *Energy Statistics Publications,*
Energy Trends, Historical Electrical Data.

In the UK, a different situation existed with the "dash-for-gas" program.[85] In 1988, Margaret Thatcher had strongly supported action on climate change. However, Bert Bolin surmised: "Her motives were, however, presumably also political. It was indeed a political advantage to have a forceful environmental justification for closing coalmines and instead developing the British oil and gas resources."[86]

With the availability of economic natural gas and the rising price of coal, a major switch between the two fossil fuels had already been underway prior to Kyoto (Figure 4.3). The British government, in fact, even slowed down this conversion by supporting their domestic coal industry.[87] This switch from coal to natural gas continues today, from which many of the future British reductions in GHGs will originate as well.[88]

Half of the emissions reductions in the entire EU came from Germany and the UK.[89] Between 1990 and 2007, EU-15 GHG emissions were reduced by 4.4 percent (807 Tg CO_{2e}), mainly through:[90]

- energy and fuel efficiency, which included East German closures – 60 percent
- switching power generation from coal (or oil) to natural gas – 27 percent.

The reduction in emissions by some members of the EU had begun well before Kyoto and had been driven purely by economics.[91] The apparent all-star performance of the EU was largely the result of lucky circumstances. But the EU deserves some credit for at least playing the carbon game, as they agreed to carbon targets and created a value for carbon through the ETS.

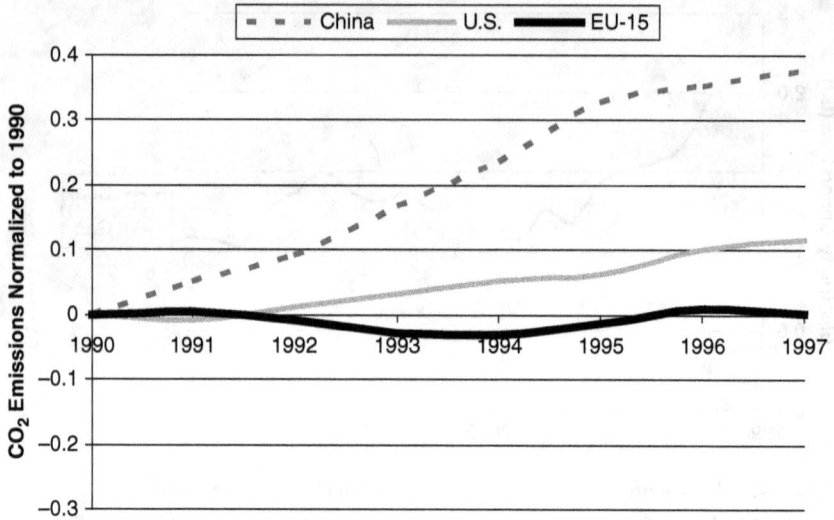

Figure 4.4. Annual CO_2 anthropogenic emissions normalized to 1990 for China, the US and the EU-15: 1990–7.

Source: BP.

Misconduct at Kyoto

Germany and the rest of the EU had almost nothing to do to meet their carbon targets, and similar constraints on other developed countries would make them less competitive, so the EU favored similar regulations for all developed countries. On the other hand, countries which had increasing emissions opposed strict carbon targets. In the US, where modest growth in emissions was expected, there was some interest in the issue but there was no support for legally binding targets. Nations with exceptional growth, such as China, were adamant that they would never agree to GHG restrictions. These national predispositions highlight a major flaw of the Kyoto process.

Strange as it may seem, the targets for Europe, including Germany, were from a practical point of view much easier to achieve than those of North America. Between 1990, the base year of Kyoto, and 1997, the year that the Kyoto Protocol was signed, no nation had deliberately established policies to reduce GHG emissions. Yet, by 1997, there was already a large disparity between China, the US and the EU-15 in terms of the change in their GHG emissions (Figure 4.4). Whereas the emissions of the US, and especially China, had increased, those of the EU were relatively unchanged. Furthermore, these trends in emissions were expected to continue. The normalized emissions rates illustrated in Figures 4.4 to 4.6 are a reliable predictive tool for determining a nation's attitude on carbon targets: as a rule, the more positive the value, the more resistance to targets,

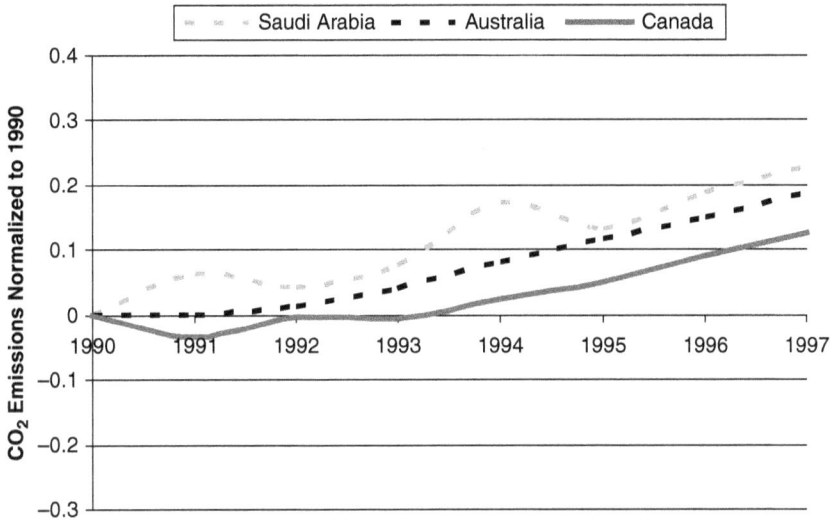

Figure 4.5. Annual CO_2 anthropogenic emissions normalized to 1990 for Saudi Arabia, Australia and Canada: 1990–7.
Source: BP.

and the more negative the value, the more the nation was an activist for strong targets for *all* (developed) nations. For example, the strongest opponents of carbon targets were China (Figure 4.4) and Saudi Arabia (Figure 4.5), which both had high rates of normalized emissions. The US, Australia and Japan were more moderate opponents, but still generally resistant to strict regulations, as would be expected given their trends in emissions. Australia was an odd case because, despite its predictable early objections, it ratified the Kyoto Protocol late into the process in 2007. The decision was especially surprising since, by this time, it was clear that it could not meet its target (average emissions from 2008 to 2012 were above 1990 by 42 percent[92] and the Kyoto target had been a rise of only 8 percent).

Was Canada an exception to the rule (Figure 4.5)? By 1997 its emissions increases were similar to that of the US, yet Canada still ratified the Kyoto Protocol at the end of 2002. But, since Canada withdrew from the Kyoto Protocol in December 2011 (as discussed earlier), Canada followed the trends-in-emissions rule after all. An interesting development took place just as the final draft of this manuscript was being prepared for publication. On June 25, 2013, President Obama stated (again) that the US would reduce GHG emissions by 17 percent by 2020 compared to 2005.[93] The announcement was made only a few weeks after the US 2012 emissions figures came out and, were already below 2005 levels by 11 percent.[94] The significance to Canada is that the government would be under pressure to match the US initiative, even though Canada's

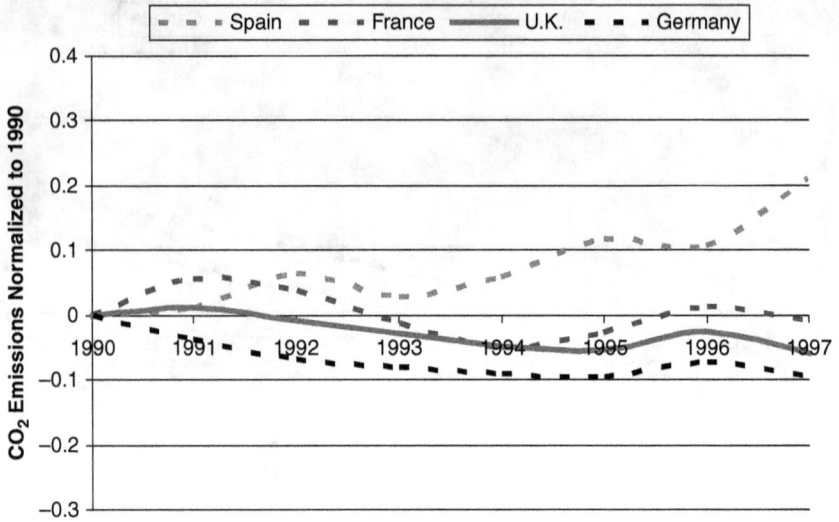

Figure 4.6. Annual CO_2 anthropogenic emissions normalized to 1990 for selected nations in the EU-15: 1990–7.

Source: BP.

2012 emissions were only 2.5 percent below 2005. Under these conditions, Canada's competitive position compared to the US would suffer if it accepted similar targets.

At first glance, another apparent exception to the rule was Spain, which had emissions growing at a faster rate than the US and Canada (Figure 4.6) but ratified the Kyoto Protocol as well. Along with Greece, Portugal and Ireland, Spain had, at first, opposed strict binding targets[95] but had been bribed to join the EU coalition with the burden-sharing[96] (also known as the "bubble") agreement which assigned it an increase in emissions of 15 percent for its Kyoto target! To compensate for the increase in emissions targets of Spain and other EU members, Germany agreed to an increased target under the burden-sharing agreement: an impressive sounding reduction of 21 percent in carbon emissions over the Kyoto base year of 1990. Germany recognized the needs of its partners with the burden-sharing agreement, but those outside of the EU had to fend for themselves. Wirth had questioned the fairness of the burden-sharing agreement on October 15, 1997,[97] and Eizenstat proposed that non-EU members of Annex I be allowed to form a similar "bubble" agreement[98] with no success.

Let's study the CO_2 emissions compared to 1990 (the base year) at critical dates in the development of the Kyoto Protocol:

- 1995: Latest year of CO_2 data when the Kyoto agreement was reached in 1997.
- 2000: Latest year of CO_2 data when the EU ratified Kyoto in 2002.

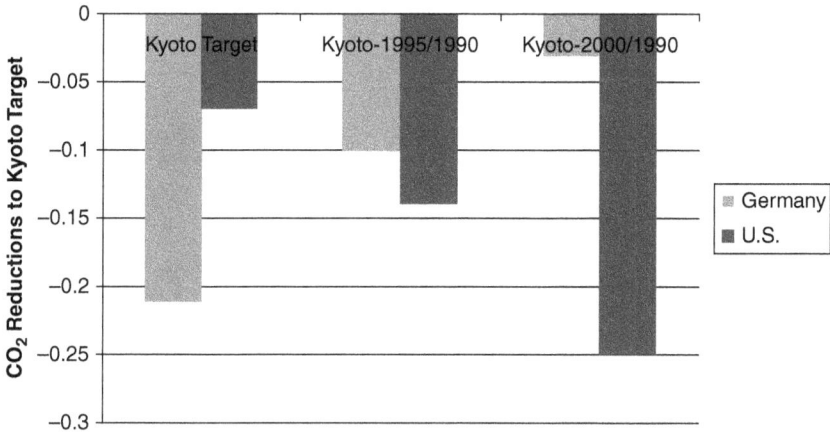

Figure 4.7. GHG emissions reductions for Germany and the US to meet their Kyoto targets.
Source: CDIAC.
Note: "Kyoto Target" is the target emissions reduction in the Protocol (column 4 of Table 4.9). "Kyoto-1995/1990" (column 4 − column 1 of Table 4.9) is the emissions reduction required to meet the Kyoto target from the actual emissions in 1995; "Kyoto-2000/1990" (column 4 − column 2 of Table 4.9) is the emissions reduction required to meet the Kyoto target from the actual emissions in 2000.

By the time of the final Kyoto negotiations the EU already knew that its 1995 emissions were below the 1990 baseline by 2 percent[99] and more declines were coming even without any GHG emissions reduction policies. Yet the EU had a Kyoto target (-8 percent) only 1 percent higher than the US (-7 percent). Also, by this time, even Germany's target was really not as challenging as the US (Figure 4.7 – middle bars). More importantly, economic conditions indicated continued lower emissions in Germany and increased emissions in the United States. In other words, despite Germany's impressive target, the potential economic impact of the Kyoto Protocol would have been much more costly for the US than Germany (and other EU member states).

Even with its more advantageous international position, intense negotiations still took place among EU members after Kyoto and ratification did not take place until five years later. Let's look at the emissions picture at the time of ratification (Figure 4.7 – third set of bars), using the available data from 2000. The competitive imbalance between the emissions of Germany and the US had just increased over the five-year period. Germany was now only 3 percent away from its Kyoto target, while the growing economy of the US had its emissions now a staggering 25 percent away.

Germany and the EU cannot be faulted for having easy emissions reductions. However, they can be criticized for deliberately attempting to force similar emissions reductions upon other developed nations that were not in such fortunate

Table 4.9. Carbon statistics (annual CO_2 emissions compared to 1990 – Kyoto base year) for selected nations.

	CO_2 Emissions			*Kyoto Target**
	1995/1990	*2000/1990*	*2010/1990*	*(2012/1990)*
Germany	–11%	–18%	–25%	–21%
UK	–1%	–5%	–14%	–13%
France	–1%	–8%	–9%	0
Italy	+3%	+5%	–4%	–7%
Spain	+13%	+29%	+21%	+15%
US	+7%	+18%	+13%	–7%
Canada	+2%	+19%	+15%	–6%

* Kyoto Target is based on the average of 2008–12 emissions.
Source: CDIAC.

positions. We can trace the EU climate change policy development back to just after the Villach Conference:

- September 12, 1986 – the EU passed a resolution to develop policies to reduce CO_2 emissions.[100]
- November 16, 1988 – the committee reported that the EU should take a leadership role in the development of policy to fight climate change;[101] Germany,[102] France and the Netherlands pushed for a united EU climate change policy, which was accepted in June 1989.
- October 29, 1990 – the first Joint Energy/Environment Council of the EU proposed that they would return to 1990 levels by 2000 *if other nations agreed to similar regulations.*[103] But the EU only had to follow a business-as-usual approach to get back to its announced targets of reaching 1990 levels by 2000.[104]

The damning admission was that the EU would only agree to carbon targets if other nations agreed to similar regulation. It continued to fight for this concession from its competitors, and an "undifferentiated target" was one of the three pillars of its negotiating strategy at Kyoto:[105]

> The European Union was very active in the international climate negotiations that led to the Kyoto Protocol. Three features characterised the European Union's negotiating position: a commitment to mandatory caps on emissions by developed countries, an undifferentiated target of 15% below 1990 emissions levels, and an antipathy toward emissions trading as a mechanism for achieving this target. The Kyoto Protocol was signed in December 1997. Signatories agreed to caps, but the EU failed to achieve its 15% reduction or undifferentiated target goal.

Berlin, the location for the first COP in 1995, was where the EU really launched its strategy, which forever changed the carbon game. Germany had become the *de facto* leader[106] of the EU on climate change. By the time the COP process began in Berlin, Germany had upped the ante by supporting high targets of 20 percent emissions reductions by 2005.[107] The meeting was under the direction of Angela Merkel (1954–), then Environment Minister for Germany. The negotiations turned into a marathon session. A senior ministerial meeting was called at 11 p.m. on April 6 which went through the night. In the morning of the following day, Merkel divided the delegates. The developed countries were cloistered in one room and the developing in another. The German Environment Minister had given shuttle diplomacy a new meaning. Merkel went back and forth between the two groups where she presented and negotiated each group's case to the other until an agreement was reached at 6 a.m. on the morning of April 8.[108] So the Berlin Mandate cleared the path to Kyoto, but at what price? No sooner did the Berlin Wall come down than a new one went up, this time between the US and Kyoto. The first Berlin Wall stood for 28 years; how long will it take for the second one to fall?

The German goal appeared to be, above all, to get a deal in Berlin, and that meant letting the developing nations off the hook, for without their support there would be no Protocol. By doing so, Germany now had the support of the developing nations. Almost overnight, the German strategy was hailed by most of the world, and any developed nation that challenged the carbon targets was greeted with howls of protest from across the globe. The Americans were trapped.

Of course, the wall fall was not a state secret. For example, at the House of Representatives meeting in May 1995, just after COP 1 at Berlin, the ramifications of the unification of Germany on emissions were freely discussed. At this hearing, John Dingell stated:[109]

> [T]he accomplishments of the German Federal Republic stands from the fact that they are now shutting down a huge amount of inefficient and energy inefficient industries in the former Eastern Germany, and they can achieve enormous efficiencies, take enormous credit for that progress by that simple fact.

Later Ron Klink (1951–), a Representative of Pennsylvania,[110] described the German emissions reductions thus: "This is all hokey. This is a fake." And a memo approved by Wirth[111] three months before Kyoto reported:

> (… Germany, with enormous emissions reductions that have accrued from closing down inefficient former East German power plants and factories), the Europeans have proposed extremely aggressive next steps. The EU has called for a 15 percent reduction in greenhouse gas emissions below 1990 levels by the year 2010 for developed countries.

The Americans even suspected that at COP 1 the Germans would give the developing countries no targets in return for their support. As discussed earlier, the Global Climate Coalition (GCC), which comprised some of the largest corporations in

America, warned the Clinton Administration that Germany and the EU were seeking a competitive advantage over the US:[112]

> Let me stress that nobody should be deceived into thinking that the Europeans' objectives are altruistic or that they are unaware that their proposals will gain them trade competitive advantage to the detriment of the United States. It also makes no sense to begin negotiations until the U.S. delegation has gained, at least in broad principle, the understanding that such negotiations must result in meaningful new commitments by developing nations ... Our fear is that in order to placate Germany and others, the developed nations in Berlin will let developing countries off the hook so that the forthcoming session can produce a Berlin accord.

And, at hearings in Congress, Frank Pallone (1951–), the Representative from New Jersey, expressed similar fears:[113]

> particularly Germany, I guess, as trying to move us in the direction of some sort of mandates – give maybe Germany and the European Union countries some sort of competitive advantage ... Is there a competitive advantage for them, and is that why they are pushing for some of these things, particularly Germany?

The US then was well aware of the advantageous position of Germany and what might happen in Berlin. Complicating matters, this intrigue was all taking place during the Clinton–Gore Administration, which wanted Kyoto to succeed. One State Department official called the EU strategy "political genius."[114] If the Americans did not submit to the mandate, they risked being vilified for destroying the negotiations. There was immense public pressure on the Clinton Administration not to be seen as "blowing up Kyoto"[115] or causing the "train wreck."[116] So the Americans signed on but they would never be players in the Kyoto season.

Later, after COP 3 in June 1998, the EU would boast about the success of its Kyoto mission:[117]

> The EU succeeded in meeting a number of its negotiating goals at Kyoto, in particular the acceptance of legally-binding targets by the EU's main competitors and trading partners similar to the EU's own commitment. This ensured that Community competitiveness, a major concern, was safeguarded.

This was, without doubt, a European victory. However, enhancing competitiveness at the expense of international cooperation, in the face of a global crisis, is unsportsmanlike and contrary to the spirit of the UNFCCC mandate.

Out of these Machiavellian climate negotiations emerged the famous Kyoto Protocol. Almost 200 nations had signed and ratified the Kyoto agreement. With such widespread global support, one might get the impression that most of the world was diligently working to fight climate change, but this was not the case.

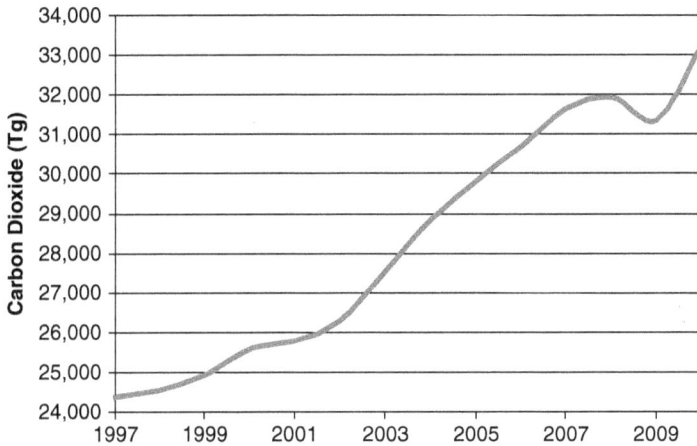

Figure 4.8. Annual CO_2 anthropogenic emissions in the world: 1997–2012.
Source: BP.

Simply ratifying the treaty did not mean emissions reductions; the problem with Kyoto was that the carbon game had many fans but not many players. About 150 of the signed nations, so-called developing nations, did not have to do anything outside of administrative duties. Among the group that actually had to reduce GHG emissions, only five (if the US and Canada are included) were non-European members, and they all had double digit increases in the average of their CO_2 emissions from 2008 to 2012 compared to 1990.[118]

Clearly, Kyoto has been essential in promoting the climate change issue, especially through the UNFCCC.[119] The Kyoto Protocol, though, must ultimately be judged by the emissions themselves. Since the Kyoto Protocol in 1997, when the world agreed to take action on climate change, global anthropogenic CO_2 emissions have soared by 40 percent (Figure 4.8).

Shortly before Cancun, Yu Qingtai, who had been the Chinese chief climate negotiator, bluntly stated: "The United Nations Framework Convention on Climate Change (UNFCCC) has been in place for almost two decades, and it has achieved next to nothing."[120]

Similar comments came, surprisingly, from the esteemed Bert Bolin, the first Chairman of the IPCC, who wrote:[121]

> The Kyoto conference did not achieve much with regard to limiting the build up of greenhouse gases in the atmosphere …
>
> The situation in many regards remains the same or has even worsened as we enter the twenty-first century. And I feel that a more pessimistic view of the future is now emerging, since progress in dealing with the problem as sketched has been slow or non-existent.

The first decade of the new millennium has witnessed by far the largest growth in GHG emissions in the history of the world. With this Kyoto season ending, and another set to begin, it is time to change how we play the carbon game.

References[122]

Benedict, R.E., Morals and Myths: A Commentary on Global Climate Policy, *WZB, Mitteilungen, Heft* 109, September 15, 2005; http://www.cba.com.hr/docs/waas/wzb_climate_policy_0905.pdf

Bodnar, P., The Political Economy of Climate Change: Did States Defend Their Material Interests When Negotiating the Kyoto Protocol, *Proceedings 2004 Annual Meeting of the American Political Science Association*, September 2004; http://www.allacademic.com/meta/p_mla_apa_research_citation/0/5/9/9/5/p59951_index.html

Bohme, S., and Dabhi, S., (eds.), *Upsetting the Offset: The Political Economy of Carbon Markets*, Mayflybooks, 2009; http://www.thecornerhouse.org.uk/sites/thecornerhouse.org.uk/files/UpsettingtheOffset.pdf

Bolin, B., *A History of the Science and Politics of Climate Change*, Cambridge University Press, 2007

Boyce, J.K., and Riddle, M., *Cap and Dividend: How to Curb Global Warming While Protecting the Incomes of American Families*, 2007; http://www.capanddividend.org/files/WP150.pdf

BP, *Statistical Review of World Energy*, 2013 (and earlier years); http://www.bp.com/en/global/corporate/about-bp/statistical-review-of-world-energy-2013.html

Buchner, B.K., and Dall'Olio, S., Russia: The Long Road to Ratification, December 2004; http://papers.ssrn.com/sol3/papers.cfm?abstract_id=627430

Cass, L.R., *The Failures of American and European Climate Policy*, SUNY, 2006; http://books.google.ca/books?id=tlrdtIz1w5wC&pg=PA33&lpg=PA33&dq=Those+who+think+we%E2%80%99re+powerless+to+do+anything+about+the+%E2%80%98greenhouse+effect%E2%80%99+are+forgetting+about+the+%E2%80%98White+House+effect.%E2%80%99&source=bl&ots=1STRiGZr62&sig=tb1_GiJSf1Ua5U2zuQ7iV5EKuSA&hl=en&ei=-m2nTNaVDIjCsAOistHuDA&sa=X&oi=book_result&ct=result&resnum=1&ved=0CBQQ6AEwAA#v=onepage&q=Those%20who%20think%20we%E2%80%99re%20powerless%20to%20do%20anything%20about%20the%20%E2%80%98greenhouse%20effect%E2%80%99%20are%20forgetting%20about%20the%20%E2%80%98White%20House%20effect.%E2%80%99&f=false

CDIAC (Carbon Dioxide Information Analysis Center) – Oak Ridge National Laboratories – US Department of Energy; data taken on May 12, 2013; (note that historic data is periodically updated by CDIAC); http://cdiac.ornl.gov/trends/emis/meth_reg.html

Climate Institute, *G20 Low Carbon Competitiveness*, February 2013; http://www.vivideconomics.com/index.php/publications/g20-low-carbon-competitiveness-index---2013-update

Cole, D.H., A Glimmer of Hope, The EU's Emissions Trading Scheme, Chapter 3, 2009; http://papers.ssrn.com/sol3/papers.cfm?abstract_id=1533566

Convery, F., Ellerman, D., and de Pertius, C., The European Carbon Market in Action: Lessons from the First Trading Period, March 2008; http://web.mit.edu/globalchange/www/ECMreport.html

Dayton, K., Germany Plans to Phase Out Coal, WyoFile, December 7, 2012; http://wyofile. com/2012/12/18592/

Deutsche Bank, *The German Feed-In Tarriff for PV*, May 2011; http://www.dbcca.com/ dbcca/EN/_media/German_FIT_for_PV.pdf

Driesen, D.M., Sustainable Development and Market Liberalism Shotgun Wedding: Emissions Trading Under the Kyoto Protocol, 2007; http://papers.ssrn.com/sol3/papers. cfm?abstract_id=986145

ECOFYS (for IEA), de Jager, D., and Rathmann, M., *Policy Instrument Design to Reduce Financing Costs in Renewable Energy Technology Projects*, October 2008; http://www. ecofys.com/files/files/report_policy_instrument_design_to_reduce_financing_costs_in_ renewable_energy_technology_pro.pdf

Earth Negotiations Bulletin; http://www.iisd.ca/vol12/

ECOFYS (for IEA), Gilbert, A., Bode J-W., and Phylipsen, D., *Analysis of the National Allocation Plans for the EU Emissions Trading Scheme*, August 2004, http://www. ecofys.com/en/publications/43/

ECOFYS (for WWF), *G8 Climate Scorecard, 2009*; http://www.ecofys.com/en/publication/ g8-climate-scorecards/

ECOFYS/McKinsey, *EU ETS Review: Report on International Competitiveness*, December 2006; http://ec.europa.eu/environment/climat/emission/pdf/etsreview/061222compreport. pdf

EEA, *EU Greenhouse Gases in 2011; More Countries on Track to meet Kyoto Targets. Emissions Fall 2.5%*, November 7, 2012; http://www.eea.europa.eu/pressroom/ newsreleases/eu-greenhouse-gases-in-2011.5

EEA, *Tracking Progress towards Kyoto and 2020 in Europe*, No. 7/2010, 2010; http:// www.eea.europa.eu/publications/progress-towards-kyoto

EEA Report, *Annual European Union Greenhouse Gas Inventory 1990 to 2009 and Inventory Report 2011*, No. 2/2011, May 27, 2011; http://www.eea.europa.eu/ publications/european-union-greenhouse-gas-inventory-2011

EEA Report, *Application of the Emissions Trading Directive by EU Member States*, No. 13/2008, 2008; http://www.eea.europa.eu/publications/technical_report_2008_13

EEA Report, *Greenhouse Gas Emission Trends and Projections in Europe 2012*, No. 6/2012, October 24, 2012; http://www.eea.europa.eu/publications/ghg-trends-and-projections-2012

EEA Report, *Greenhouse Gas Emission Trends and Projections in Europe 2011*, No. 4/2011, October 7, 2011; http://acm.eionet.europa.eu/reports/EEA_Rep_4_2011_GHG_ trend_ projections

EEA Report, *Greenhouse Gas Emissions Trends and Projections in Europe 2009*, No. 9/2009, 2009; http://www.eea.europa.eu/publications/eea_report_2009_9

Ellerman, A.D., and Buchner, B., *Over-Allocation or Abatement, MIT Program on the Science and Policy of Global Change*, Report No. 141, December 2006; http://dspace. mit.edu/bitstream/handle/1721.1/35837/MITJPSPGC_Rpt141.pdf?sequence=1

Ellerman, A.D., and Joskow, P.L., The European Union's Emissions Trading Scheme in Perspective, May 2008; http://www.pewclimate.org/docUploads/EU-ETS-In-Perspective-Report.pdf

Ellerman, A.D., Convery, F.J., and de Perthuis, C., *Pricing Carbon: The European Union Emissions Trading System*, Cambridge University Press, 2010; http://books.google. com/books?id=pigVprvX56QC&lpg=PP1&ots=CrZE2aIcRB&dq=%22Pricing%20 Carbon%22%20ellerman&pg=PP1#v=onepage&q&f=false

Environmental Defense Fund, *The World's Carbon Markets*, May 2013; http://www.edf. org/climate/worlds-carbon-markets

Ernst & Young, *Renewable Energy Country Attractiveness Indices*; http://www.ey.com/GL/en/Industries/Oil ---Gas/Oil_Gas_Renewable_Energy_Attractiveness-Indices; accessed May 2013

European Commission, *A Community Strategy to Limit Carbon Dioxide Emissions and to Improve Energy Efficiency*, Sec (91) 1744, October 1991; http://aei.pitt.edu/4931/

European Commission, *A Roadmap for Moving to a Competitive Low Carbon Economy in 2050*, Communication, March 2011; http://ec.europa.eu/clima/policies/roadmap/documentation_en.htm

European Commission, *Building a Global Carbon Market*, November 2006; http://ec.europa.eu/environment/climat/emission/pdf/com2006_676final_en.pdf

European Commission, *Climate Action, Policies, Emissions Trading System*, July 1, 2013; http://ec.europa.eu/clima/policies/ets/index_en.htm

European Commission, *Climate Change – Towards an EU Post-Kyoto Strategy*, Communication, June 1998; http://aei.pitt.edu/6815/01/003692_1.pdf; http://aei.pitt.edu/6815/; accessed November 2011

European Commission, *Energy for the Future: Renewable Sources of Energy*, Communication, November 1997; http://europa.eu/documents/comm/white_papers/pdf/com97_599_en.pdf

European Commission, *Europe 2020*, Communication, March 2010; http://eur-lex.europa.eu/LexUriServ/LexUriServ.do?uri=COM:2010:2020:FIN:EN:PDF

European Commission, *Preparing for Implementation of the Kyoto Protocol*, COM (1999) 230, May 1999; http://ec.europa.eu/environment/docum/pdf/99230_en.pdf; accessed November 2011

European Commission, *Questions and Answers on the Revised EU Emissions Trading System*, Memo-08–796, December 17, 2008; http://europa.eu/rapid/pressReleasesAction.do?reference=MEMO/08/796

European Commission, *Renewable Energy Roadmap*, Communication, January 2007; http://eur-lex.europa.eu/LexUriServ/LexUriServ.do?uri=CELEX:52006DC0848:EN:NOT

European Commission, *State Aid: Commission Opens In-depth Investigation into Subsidies for Producers of Non-ferrous Metals in Germany*, Press Releases, November 17, 2010; http://europa.eu/rapid/pressReleasesAction.do?reference=IP/10/1520&format=HTML

European Commission, *The Energy Dimension of Climate Change*, Communication, May 1997; http://aei.pitt.edu/4723/01/000817_1.pdf

European Commission, *The Greenhouse Effect and the Community*, Communication, November 1988; http://aei.pitt.edu/5684/01/003076_1.pdf; accessed November 2011

European Commission, *The Renewable Energy Progress Report*, Communication, April 2009; http://eur-lex.europa.eu/LexUriServ/LexUriServ.do?uri=COM:2009:0192:FIN:EN:PDF

European Commission, *The Share of Renewable Energy in the EU*, Communication, May 2004; http://eur-lex.europa.eu/LexUriServ/LexUriServ.do?uri=COM:2004:0366:FIN:EN:PDF

European Council Directive, 93/76/EEC, September 1993; http://eur-lex.europa.eu/LexUriServ/LexUriServ.do?uri=CELEX:31993L0076:EN:HTML

European Energy Exchange; http://www.eex.com/en/Market%20Data/Trading%20Data/Emission%20Rights

European Renewable Energy Council, Organization; http://www.erec.org/organisation.html

Gilbertson, T., and Reyes, O., Carbon Trading: How It Works and Why It Fails, *Critical Currents* (7), November 2007; http://tni.org/sites/www.tni.org/files/download/carbon-trading-booklet.pdf

Gipe, P., *2012* German Nuclear and Gas-fired Generation Falls Further While Renewables Grow, *Renewable Energy World*, March 19, 2013; http://www.renewableenergyworld.com/rea/news/article/2013/03/2012-german-nuclear-gas-fired-generation-falls-further-while-renewables-grow?cmpid=rss&utm_source=twitterfeed&utm_medium=twitter

Grubb, M., Brewer, T.L., Sato, M., Heilmayr, R., and Fazekas, D., Climate Policy and Industrial Competitiveness: Ten Insights from Europe on the EU Emissions Trading System, August 2009; http://www.climatestrategies.org/component/reports/category/61/204.html

Hansen, J., Climate change Time Bomb: Actions Needed to Avert Disaster, October 26, 2009; http://www.columbia.edu/~jeh1/

Harvey, F., Europe's Climate Change Chief Vows to Fight to Save Emissions Trading Scheme, *Guardian*, April 17, 2013; http://www.guardian.co.uk/environment/2013/apr/17/europe-climate-chief-vow-save-emissions-trading

Harvey, F., UK Green Energy Projects Fall by the Wayside in Dash for Gas, *Guardian*, December 4, 2011; http://www.guardian.co.uk/environment/2011/dec/04/renewableenergy-energy-industry?newsfeed=true

Hecht, M.T., and Tirpak, D., Framework Agreement on Climate Change: A Scientific and Policy History, *Climatic Change* 29 (4), 371, 1995

Helm, D., To Slow Warming, Carbon Tax, *New York Times*, November 11, 2012; http://www.nytimes.com/2012/11/12/opinion/on-climate-change-the-us-is-doing-better-than-europe.html?partner=rssnyt&emc=rss&_r=0; accessed November 2012

Hepburn, C., Carbon Trading: A Review of the Kyoto Mechanisms, *Annual Review of Environment and Resources*, 32, 375, 2007

IEA, Germany to Win Coal Aid Deal, *Energy Efficiency and Climate Change News*, December 8–14, 2010; http://www.iea.org/newsletters/ee_cc/index.asp; accessed November 2011

IEA, Statistics: CO_2 *Emissions from Fuel Combustion, Highlights – 2011*; http://www.iea.org/publications/free_new_Desc.asp?PUBS_ID=2450; accessed November 2011

IEA/OECD, Reinaud, J., and Philibert, C., *Emissions Trading: Trends and Prospects*, December 2007; http://www.iea.org/publications/free_new_Desc.asp?PUBS_ID=2001; accessed November 2011

International Emissions Trading Association; http://www.ieta.org/

Lang, M., and Mutschler, U., Renewable Energy Source Most Important Electrical Source in Germany in 2013?, German Energy Blog, August 13, 2012; http://www.germanenergyblog.de/?p=10070

Leslie, K., Ontario Municipalities to Get More Say on Wind Farms, and Chance to become Partners, *Globe and Mail*, May 30, 2013; http://m.theglobeandmail.com/news/national/ontario-municipalities-to-get-more-say-on-wind-farms-and-chance-to-become-partners/article12264458/?service=mobile

Macalister, T., Gas Strategy Unveiled by George Osborne, *Guardian*, December 5, 2012; http://www.guardian.co.uk/environment/2012/dec/05/gas-strategy-unveiled-george-osborne

Maher, S., Europe's $287BnCarbon 'Waste': UBS Report, *The Australian*, November 23, 2011; http://www.theaustralian.com.au/national-affairs/europes-287bn-carbon-waste-ubs-report/story-fn59niix-1226203068972

Mendonca, M., *Feed-In-Tariffs*, Earthscan, May 1, 2007; http://books.google.ca/books?id
=TJlBl8eFpTkC&pg=PA27&lpg=PA27&dq=stromeinspeisungsgesetz&source=bl&ots
=Jp1MMTo49T&sig=_k7K4rM3RYq2LpVw7sXwysZaIxQ&hl=en&ei=R9mnS5l1kO
qxA_-5uYoE&sa=X&oi=book_result&ct=result&resnum=8&ved=0CCEQ6AEwBw#v
=onepage&q=stromeinspeisungsgesetz&f=false

National Round Table on the Environment and the Economy, *Climate Prosperity – Report #1*, *Measuring Up: Benchmarking Canada's Competitiveness in a Low Carbon World*, May 20, 2010; http://nrtee-trnee.ca/climate/climate-prosperity/measuring-up-advise; accessed November 2011

Nicolosi, M., and Fuersch, M., Implications of the European Renewables Directive on RES-E Support Scheme Designs and its Impact on the Conventional Power Markets, *International Association of Energy Economics*, Q3, 2009; http://www.iaee.org/documents/newsletterarticles/309nicolosi.pdf

Official Journal of the European Communities, DIRECTIVE 2009/28/EC, April 23, 2009; http://eur-lex.europa.eu/LexUriServ/LexUriServ.do?uri=OJ:L:2009:140:0016:0062 :EN:PDF; also see Council of the European Union, *Council Adopts Climate-Energy Legislative Package*, April 6, 2009

OPA, *Renewable Energy Feed-in-Tariff Program*, *FIT Price Schedule*; http://fit. powerauthority.on.ca/program-archives; accessed November 2011

P5_TA-PROV (2004) 0276; *European Parliament Resolution on the International Conference for Renewable Energies*, June 2004; http://www.ee-netz.de/eu_targets/ep_ resolution_on_renewables01042004.pdf

Payne, S., The Great British Wind Scam: The Government Responds, *The Spectator*, November 24, 2012; http://blogs.spectator.co.uk/sebastian-payne/2012/11/how-the-government-

Reguly, E., Austerity Pulling Plug on Europe's Green Subsidies, *Globe and Mail*, January 26, 2011; http://www.theglobeandmail.com/report-on-business/commentary/eric-reguly/austerity-pulling-plug-on-europes-green-subsidies/article1883888/

Ren21, *Global Status Report*; http://www.ren21.net/REN21Activities/Publications/ GlobalStatusReport/tabid/5434/Default.aspx

Revkin, A.C., China Sustains Blunt 'You First' Message on CO_2, *New York Times*, September 2, 2010; http://dotearth.blogs.nytimes.com/2010/09/02/china-sustains-blunt-you-first-message-on-co2/

RICS Research, ZC^2 – *The 2010 Global Zero Carbon Capacity Index*, September 2011; http://www.rics.org/site/download_feed.aspx?fileID=10479; accessed November 2011

Samson , R.A., and Stamler, S.B., *Going Green for Less: Cost-Effective Alternative Energy Sources*, February 2009; http://www.cdhowe.org/pdf/commentary_282.pdf

Schiller, B., Europe's CO2 Trading Scheme: Is it Time for a Major Overhaul, *Yale Environment 360*, April 28, 2011; http://e360.yale.edu/feature/europes_co2_trading_ scheme_is_it_time_for_a_major_overhaul/2396/; accessed January 2012

de Sepibus, J., Linking the EU Emissions Trading Scheme to JI, CDM and Post-2012 Offsets, 2008; http://papers.ssrn.com/sol3/papers.cfm?abstract_id=1283523

Stern, N., Stern Review on the Economics of Climate Change, 2006; http://webarchive. nationalarchives.gov.uk/+/ http://www.hm-treasury.gov.uk/sternreview_index.htm

The Economist, ETS, RIP?, April 20, 2013; http://www.economist.com/news/finance-and-economics/21576388-failure-reform-europes-carbon-market-will-reverberate-round-world-ets

The Economist, Europe's Dirty Little Secret, January 3, 2013; http://www.economist.com/ news/briefing/21569039-europes-energy-policy-delivers-worst-all-possible-worlds-unwelcome-renaissance

Tilford, S., How to Make EU Emissions Trading A Success, May 2008; http://www.cer.org.uk/publications/archive/report/2008/how-make-eu-emissions-trading-success

UK Department of Energy and Climate Change (UK), *Energy Statistics Publications, Energy Trends, Historical Electrical Data*; https://www.gov.uk/government/statistical-data-sets/historical-electricity-data-1920-to-2011; accessed May 2013

UNFCCC, *Conference of the Parties, First Session – Part II*, Berlin, March 28, 1995; http://unfccc.int/resource/docs/cop1/07a01.pdf

UNFCCC, *Conference of the Parties, Second Session – Part I*, Geneva, October 29, 1996; http://unfccc.int/resource/docs/cop2/15.pdf

UNFCCC, *Conference of the Parties, Third Session – Part II*, Kyoto, March 25, 1998; http://unfccc.int/resource/docs/cop3/07a01.pdf

UNFCC, *International Emissions Trading*; http://unfccc.int/kyoto_protocol/mechanisms/emissions_trading/items/2731.php

US Department of Energy, National Renewable Energy Laboratory, *A Policymaker's Guide to Feed-in-Tariff Policy Design*, 2010; http://www.nrel.gov/docs/fy10osti/44849.pdf

US Department of State, *Climate Change: An Emerging Policy on Developing Country Participation, Information Memorandum*, September 4, 1997; http://www2.gwu.edu/~nsarchiv/NSAEBB/NSAEBB303/Document%206.pdf

US Department of State, OES, *Climate Change Negotiations: Key Issues Facing US in the Run-up to the Kyoto Conference*, September 5, 1997; http://www2.gwu.edu/~nsarchiv/NSAEBB/NSAEBB303/Document%205.pdf

US Department of State, *The Kyoto Meeting: A Scenario for Redefining Success*, September 24, 1997; http://www2.gwu.edu/~nsarchiv/NSAEBB/NSAEBB303/Document%2012.pdf

US House of Representatives, Hearings before the Subcommittee on Energy and Power, 104th Congress, Serial Nos. 104–13, *International Global Climate Change Negotiations*, March 21 and May 19, 1995.

US House of Representatives, Hearings before the Subcommittee on Energy and Power, 105th Congress, Serial Nos. 105–108, *The Kyoto Protocol and Its Economic Implications*, March 4, 1998; http://books.google.ca/books?id=q19o4MUr87cC&pg=PA13&dq=Stuart+Eizenstat,+kyoto&hl=en&ei=27NqTMnCJY30tgOE2rgw&sa=X&oi=book_result&ct=result&resnum=1&ved=0CCoQ6AEwAA#v=onepage&q=Stuart%20Eizenstat%2C%20kyoto&f=false

Usui, Y., New Modes of Governance and the Climate Change Strategy in the European Union, 2005; http://project.iss.u-tokyo.ac.jp/crep/pdf/ws05/3pa.pdf

Wang, U., Germany's Solar Identity Crisis, *Renewable Energy World*, January 10, 2012; http://www.renewableenergyworld.com/rea/news/article/2012/01/germanys-solar-identity-crisis?cmpid=WNL-Wednesday-January11-2012

White House, President Barack Obama, *The President's Climate Action Plan*, June 2013; http://www.whitehouse.gov/sites/default/fi les/image/president27sclimateactionplan.pdf

World Bank, ECOFYS, *Mapping Carbon Pricing Initiatives*, May 2013; http://www-wds.worldbank.org/external/default/WDSContentServer/WDSP/IB/2013/05/23/000350881_20130523172114/Rendered/PDF/779550WP0Mappi0til050290130morning0.pdf

World Bank, *State and Trends of the Carbon Market*; http://wbcarbonfinance.org/Router.cfm?Page=DocLib&ht=25621&dtype=25622&dl=0

Yale Center for Environmental Law & Policy, *Environmental Performance Index*; http://envirocenter.yale.edu/?page=environmental-performance-index

Yang, A., and Cui, Y., *Global Coal Risk Assessment: Data Analysis and Market Research*, World Resources Institute, November 2012; http://www.wri.org/publication/global-coal-risk-assessment

Notes

1 US Department of Energy, 2010, para. 4.2.1.1, p. 25.
2 Ibid., para. 4.2.1.1, p. 2.
3 A detailed review of FIT is presented in Mendonca, May 1, 2007; see Nicolosi and Fuersch, 2009, pp. 25–6.
4 A summary of some of the EU incentives is presented in ECOFYS, de Jager and Rathmann, October 2008; for Germany see p. 53; also see Stern, 2006, p. 367.
5 Cass, 2006, p. 190.
6 Ren21, *Global Status Report*, 2010, pp. 11, 37–41, 62; also see Ren21, *Global Status Report*, 2011, pp. 52–4.
7 European Renewable Energy Council, Organization: http://www.erec.org/organisation.html
8 Ren21, *Global Status Report*, 2011, p. 47; also see *Global Status Report*, 2010, p. 52.
9 For a positive report on the German program see Deutsche Bank, May 2011.
10 US Department of Energy, 2010, p. v.
11 Nicolosi and Fuersch, 2009, p. 28; Wang, January 10, 2012.
12 For an article on the unsustainable nature of FIT programs see Reguly, January 26, 2011.
13 Leslie, May 30, 2013.
14 Calculation: ($443/MWh – $135/MWh)/0.87 tonnes of CO_2 per MWh (assumes replacement of coal to produce electricity) = $354 per tonne of CO_2.
15 The FIT program in Britain has a special problem with the size of wind power projects which has been described as "a veritable license to print money"; Payne, November 24, 2012.
16 Howe, February 2009, *The Study in Brief*.
17 European Commission, November 1997.
18 European Commission, January 2007.
19 European Commission, May 2004, p. 13.
20 European Commission, January 2007, p. 8.
21 European Commission, May 2004, p. 8.
22 European Commission, *Climate Action, Policies, Emissions Trading System*, July 1, 2013; also see International Emissions Trading Association.
23 UNFCCC, *International Emissions Trading*; http://unfccc.int/kyoto_protocol/mechanisms/emissions_trading/items/2731.php
24 UNFCCC, March 25, 1998, paragraph 12, p. 18.
25 Carbon offsets allow trade of carbon credits from another nation, usually on a project basis. Theoretically, the offset system helps fight climate change, which is a global phenomenon, and the impact of GHG emissions on the world does not depend on the country of source. For a criticism of offsets, CDM and the ETS, see Bohme and Dabhi, 2009; Gilbertson and Reyes, November 2009; Driesen, 2007.
26 For a review of the early trading of the CERs from the CDM see Hepburn, 2007, p. 385, and de Sepibus, 2008.
27 European Commission, Memo-08–796, December 17, 2008, Question 1; an overview of the development of the EU ETS is presented by Cole, 2009.
28 EEA Report, No. 13/2008, p. 57.
29 The "cap and dividend" system is quite different. The allowances are auctioned and then refunded to the populace on a per capita basis, which was proposed in the US Senate in the Cantwell–Collins Bill, see Boyce and Riddle, 2007. A related but more direct system is "fee and dividend," where the cap is replaced by a carbon

tax (or fee); see Hansen, October 26, 2009. Both systems are a tax and refund per capita system.

30 http://ec.europa.eu/environment/climat/emission/auctioning_en.htm
31 Grubb *et al.*, August 2009, p. 14.
32 For a detailed review of Phase I, see Ellerman *et al.*, 2010.
33 Cole, 2009, pp. 23–4.
34 EEA Report, No. 13/2008, pp. 14–15; also see Cole, 2009; Ellerman and Buchner, 2006.
35 Grubb *et al.*, August 2009, pp. 10–11.
36 European Commission, May 2004, p. 23; also see ECOFYS, Gilbert *et al.*, August 2004, p. 4.
37 For a defense of Phase I, see Convery *et al.*, March 2008; and European Commission, November 2006.
38 IEA/OECD, Reinaud and Philibert, December 2007, p. 24.
39 Grubb *et al.*, August 2009, p. 13.
40 ECOFYS/McKinsey, December 2006, p. 20.
41 Grubb *et al.*, August 2009, p. 4; also see pp. 12, 18; for various other reports on the profiting from Phase I see Cole, 2009, pp. 31–2.
42 See Cole, 2009, p. 31; Tilford, May 2008; a defense of Phase I was presented by Ellerman and Joskow, May 2008.
43 Mg is the SI unit equivalent to one metric tonne, or one million grams.
44 A warm winter, for example, will reduce emissions, as took place in Europe in 2006 (see Cole, 2009, p. 31).
45 EEA Report, No. 4/2011, October 7, 2011, p. 46.
46 Grubb *et al.*, August 2009, p. 3.
47 European Energy Exchange.
48 See annual overviews of the carbon markets by the World Bank; also see Convery *et al.*, March 2008, p. 13; Environmental Defense Fund, May 2013.
49 World Bank, ECOFYS, *Mapping Carbon Pricing Initiatives*, May 2013, p. 18.
50 Maher, November 23, 2011; also see Schiller, April 28, 2011.
51 IEA/OECD, Reinaud and Philibert, December 2007, p. 9; Grubb *et al.*, August 2009, p. 12.
52 IEA/OECD, Reinaud and Philibert, December 2007, p. 8.
53 Grubb *et al.*, August 2009, p. 14.
54 EEA Report, No. 13/2008, p. 58; also see Grubb *et al.*, August 2009, p. 15.
55 In late 2010, Germany planned to issue rebates to the non-ferrous metal producers for their higher electricity costs; see European Commission, November 17, 2010.
56 EEA Report, No. 13/2008, p. 61; for an update see EEA Report, No. 4/2011, p. 41.
57 European Commission, Memo-08–796, Question 4.
58 European Commission, Memo-08–796, Question 5.
59 IEA/OECD, Reinaud and Philibert, December 2007, p. 24.
60 European Commission, Memo-08–796, Question 7.
61 Harvey, April 17, 2013.
62 *The Economist*, April 20, 2013.
63 BP, 2011.
64 P5_TA-PROV (2004) 0276, June 2004.
65 Nicolosi and Fuersch, 2009, p. 25.
66 Official Journal of the European Communities, DIRECTIVE 2009/28/EC, April 23, 2009; a progress report on RES was issued the following day, European Commission, April 2009; also see European Commission, March 2010.
67 See for example US Department of State, OES, September 5, 1997, paragraph 60; the problem was also raised at Kyoto, *Earth Negotiations Bulletin*, 12 (76), December 13, 1997, p. 9; see Bolin, 2007, pp. 148–9, 153.

68 See Bodnar, September 2004, p. 16 for a discussion.
69 BP, 2013.
70 Cole, 2009, p. 26.
71 Grubb *et al.*, August 2009, p. 26; Cole, 2009, p. 13; *Earth Negotiations Bulletin*, 12 (567), December 11, 2012, p. 26; for a detailed review of Russia and Kyoto see Buchner and Dall'Olio 2004, and Bolin, 2007, pp. 187–190.
72 EEA, November 7, 2012; EEA, October 24, 2012, p. 59; also see EEA, No. 7/2010, *Tracking Progress towards Kyoto and 2020 in Europe*, 2010, p. 30.
73 European Commission, March 2011, p. 14.
74 US House of Representatives, May 19, 1995, p. 141; US House of Representatives, March 4, 1998, pp. 44, 90; Grubb *et al.*, August 2009, p. 23.
75 European Commission, May 1999, p. 2; also see Bodnar, September 2004, p. 22; Cole, 2009, pp. 4–5; Benedict, September 2005, p. 16; Cass, 2006, pp. 45, 59, 192.
76 UNFCCC, March 28, 1995, paragraph 6, p. 50.
77 Cass, 2006, p. 59.
78 Ibid., pp. 107–10.
79 Ibid., pp. 139–40, 186.
80 IEA, December 8–14, 2010
81 Yang and Cui, November 2012, pp. 3, 5, 10, 59.
82 *The Economist*, January 3, 2013.
83 Lang and Mutschler, August 13, 2012; also see Dayton, December 7, 2012; and Gipe, March 19, 2013.
84 Cass, 2006, pp. 193, 204.
85 Ibid., pp. 84–6, 120.
86 Bolin, 2007, p. 57.
87 Cass, 2006, p. 179.
88 Harvey, December 4, 2011; Macalister, December 5, 2012.
89 EEA Report, No. 13/2008, p. 23.
90 EEA Report, No. 9/2009, p. 36.
91 Bodnar, September 2004, p. 23; see also Grubb *et al.*, August 2009, p. 23.
92 BP, 2013.
93 The White House, President Barack Obama, June 2013.
94 BP, 2013.
95 Bodnar, September 2004, p. 22.
96 *Earth Negotiations Bulletin*, 12 (76), December 13, 1997, p. 2; Cass, 2006, pp. 143–4; Usui, 2005, pp. 17, 30.
97 Cass, 2006, p. 131.
98 Ibid., p. 176.
99 European Commission, May 14, 1997, p. 3.
100 See European Commission, November 1988, pp. 60–1.
101 Ibid., p. 52.
102 Germany formed the Enquete Commission (Inquiry Commission on Preventive Measures to Protect the Atmosphere in December 1987); their findings were published in October 1988, *Protecting the Earth's Atmosphere: An International Challenge*; see Cass, 2006, p. 43.
103 European Commission, October 1991, p. 2; also see European Council Directive, 93/76/EEC, September 1993; and European Commission, May 1999, p. 2.
104 European Commission, May 1999, p. 2.
105 Convery *et al.*, March 2008, pp. 7–8; see European Commission, May 1997, p. 2; UNFCCC, *Conference of the Parties, Second Session – Part I*, p. 48; also see European Commission, May 2004, paragraph 109, p. 32.
106 Cass, 2006, pp. 44–5, 48.
107 US House of Representatives, March 21, 1995, p. 65.

108 *Earth Negotiations Bulletin*, 12 (21), April 10, 1995, p. 11; also see House of Representatives, May 19, 1995, p. 125.
109 US House of Representatives, May 19, 1995, p. 141.
110 US House of Representatives, March 4, 1998, p. 91.
111 US Department of State, OES, September 5, 1997, paragraph 41.
112 US House of Representatives, March 21, 1995, p. 74; also see p. 38.
113 Ibid., pp. 37–8; also see pp. 74, 98. The Global Climate Coalition repeated the concern after Berlin as well.
114 Hecht and Tirpak, 1995, p. 401.
115 US Department of State, *Information Memorandum*, September 4, 1997.
116 US Department of State, September 24, 1997.
117 European Commission, *Climate Change – Towards an EU Post-Kyoto Strategy*, June 1988, p. 4.
118 BP, 2013.
119 IEA, 2011, p. 12.
120 Revkin, September 2, 2010.
121 Bolin, 2007, p. 152; also see Helm, November 11, 2012 for similar comments.
122 All web-based references were confirmed (accessed) in June 2013, unless otherwise stated.

5   The game plan for next season

The future of the IPCC

On an international front, the IPCC has accomplished much and is indirectly responsible for the UNFCCC, the Kyoto Accord and the Bali Roadmap. Here lies the prowess of this organization and the Assessment Reports. Unfortunately, the IPCC has failed dismally on the more important national level. An argument can be made that the IPCC never influenced any government; those that had already determined to act on climate change, such as Germany, embraced the IPCC recommendations, and those that rejected any policy initiatives, such as the US, simply ignored them.[1] A commentary in *Nature* stated:[2]

> The crucial point here is that no amount of reform of the IPCC, or rooting out of bad science – or of scientists behaving badly – will begin to correct the flaws in the dominant approach to climate policy. Rehabilitation of climate policy is a matter not of getting the science right, but of getting the politics right.

With psychological, social and especially economic barriers, divisive politics, an entertainment-driven tabloid news media and the continuous blitz of climate change skeptics, can the IPCC ever influence public opinion in America? Should the AR5, then, be the last Assessment Report? As Benjamin Franklin aptly stated: "The definition of insanity is doing the same thing over and over and expecting different results."

The single greatest weakness of the IPCC is that the Assessment Reports have never caught the imagination of the general populace; therein rests the cause of its failure on a national level. The ability of the IPCC to influence public opinion has long been a concern; as early as the FAR, the IPCC had identified the importance of this goal.[3] An inherent barrier to the Assessment Reports, even the Synthesis Reports, is that they are scientific reviews. Science is not public friendly, and climate science especially requires a high degree of scientific literacy. By their nature, the Assessment Reports are "dry" and, at worst, devoid of meaning for the average person. The climate change picture needs to be "framed"[4] to be meaningful to the masses.[5]

We cannot be critical of the IPCC in this regard because, clearly, the influence of an Assessment Report is restricted by the policy neutral rule. Under this constraint, climate scientists can list the options to mitigate climate change but not prioritize them. Politicians may decide, in consideration of legitimate reasons, to ignore some recommendations; however, they should not prevent recommendations from even being made, which is precisely what is happening with climate change. A natural question is why does the policy neutral rule exist? An obvious answer is so that politicians will not have to defend their actions, or lack thereof, on the recommendations of the world's leading experts. The policy neutral rule protects the former from the latter by keeping the public in the dark on what experts think should be done to fight climate change. Since governments do not react to science – they react to the electorate – scientific evidence must sway public opinion to stimulate political will (Figure 5.1). The full potential of the IPCC must be unleashed;[6] politicians should not fear policy recommendations by the leading scholars of the world. As Tolba had told the IPCC in its early years, the organization "should bravely inform the world what ought to be done."[7]

The strategy

We have exposed the inequity of the Kyoto targets. Acknowledging that two decades of international negotiations have failed is not easy, but it is time to move on. Especially with the first phase of Kyoto having ended in 2012, the opportunity exists for trying something new. If we hope to win the next season, we must start now, as the lauded *Stern Review* warned:[8]

> From all of these perspectives, the evidence gathered by the Review leads to a simple conclusion: the benefits of strong, early action considerably outweigh the costs.
>
> The evidence shows that ignoring climate change will eventually damage economic growth. Our actions over the coming few decades could create risks of major disruption to economic and social activity, later in this century and in the next, on a scale similar to those associated with the great wars and the economic depression of the first half of the 20th century.

Stern compared the economic threat of climate change to the Depression and WWI and WWII. By doing so, he also, unintentionally, mentioned solutions to climate change. History has shown that major economic downturns and wars are an effective but painful "cure" for the climate change problem (Table 5.1). The largest percentage and absolute drop in industrial CO_2 emissions took place between 1929 and 1933, when emissions dropped by more than 1,000 Tg. This record would have been broken in 2009 during the Great Recession except for the huge increase in emissions from China.

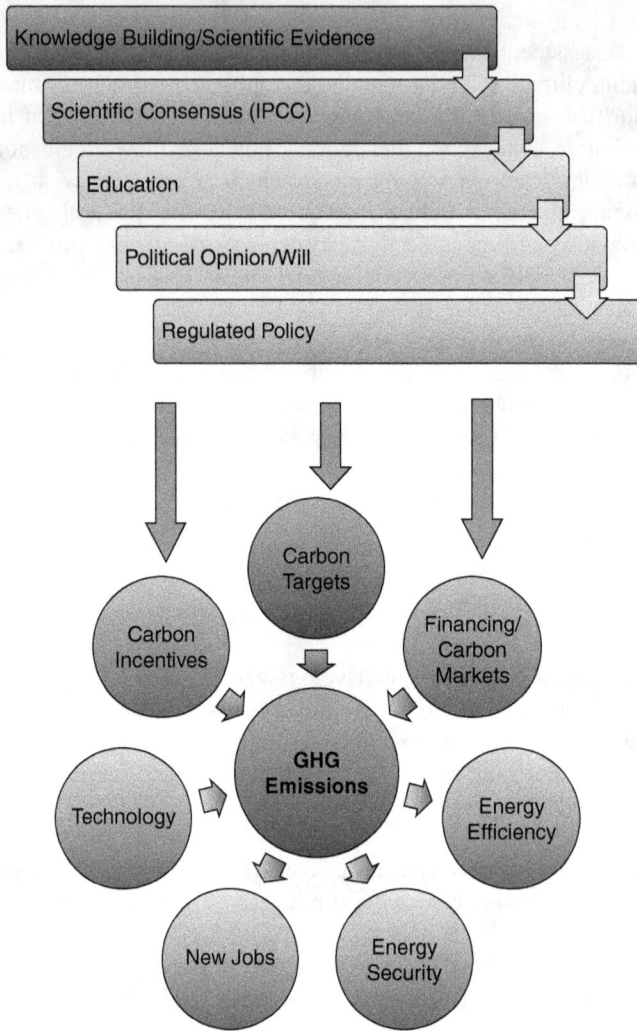

Figure 5.1. The mechanism of climate policy development.

While the UNFCCC is to be commended for its efforts to combat climate change, the process, itself, of requiring almost 200 nations to agree (to anything), is doomed. If towns, cities and nations cannot reach consensus on climate change, how can the world? We must seek an approach which does not depend only on a single binding global treaty, where most of the nations don't reduce emissions anyway. Even the former UN climate chief, Yver de Boer, has stated that another "binding international deal" is not the way forward.[9]

Table 5.1. Periods of major declines in anthropogenic CO_2 emissions.

Era	Years	CO_2 Emissions	
WWI	1917–19	–16%	–546 Tg
Great Depression	1929–33	–26%	–1,093 Tg
WWII	1943–5	–17%	–847 Tg
"Reagan Recession"	1980–3	–4%	–796 Tg
"Great Recession"	2008–9	–1%	–411 Tg

Source: Data from CDIAC.

A series of treaties to determine carbon targets should be established where only the direct nations involved would have a vote, and where each treaty would build on its predecessor:[10]

1. COP: in consultaion with the IPCC would set a long-term global GHG emissions target.[11] A target date of 2050,[12] or later, would work. Since the 1990 base year of Kyoto is far behind us, a base year of 2010 is reasonable, or, better still, an average over a five-year period.
2. "G-3" (US, EU, China): global leadership would be undertaken by the major emitters first. When over 50 percent of global GHG emissions come from three political entities, why compound the issue at the initial stage of this crisis by looking for consensus from 200 nations? The carbon triumvirate must come up with differentiated targets for themselves that will establish a model for other nations to follow. If they cannot do this, adding more nations into the mix will not help. Agreement would produce the critical mass to do something meaningful with regard to climate change and provide political impetus for other nations to join.
3. G-20: afterwards, the remaining G-20 members would be brought into the discussions.
4. COP: only then would the full COP become involved. Beyond the G20 targets already defined, the remaining G77 and other nations would have minimal responsibilities, unless there was a significant change in their economic status. Emissions targets would be on a GDP threshold basis for this group.

Climate change advocates can sometimes make a mockery of the real political and economic challenges that face governments by making it sound easy to implement policies to reduce GHG emissions – if Europe can do it, why can't you? Europe is just more fortunate that its economic path happened to reduce its GHG emissions as a bonus. Both Europe and the US make policies based on economics.[13] As Upton Sinclair (1878–1968) wrote in 1935:[14] "It is difficult to get a man to understand something, when his salary depends upon his not understanding it!" The new negotiations on differentiated targets must consider that there are a number of intrinsic energy factors which influence the economic cost to reduce GHGs:

1. *GDP & Energy Growth* – energy growth is tied to economic activity, and GHG emissions increase in a growing economy. We have reached a climate change crisis because of the economic prosperity of the industrialized world over the past century, and the most serious threat going forward is the increasing prosperity of the emerging nations, such as India, Brazil and especially China. Any country experiencing higher economic growth will have higher emissions growth, until we find a way to decouple the two. Countries with a declining economy will automatically have lower emissions.

2. *Natural Resources & Energy Security* – countries with coal reserves, such as China (or petroleum, in the case of OPEC nations), will exploit their indigenous resources. The building of a coal-fired power plant on top of a coal reserve makes economic and political sense. These plants produce electricity at the lowest cost, provide employment for the production of the electricity and mining of the coal and are a source of energy security. Conversely, a lack of such resources is an incentive to support climate change legislation as jobs and energy security will be created from a new renewable energy sector.

3. *Electrical Infrastructure & Energy Costs* – most public utility facilities were installed decades before the threat of climate change was known; some countries/regions installed hydro, some nuclear and others coal. The GHG emissions regulations are much more economically challenging for the coal-based producers, especially if their coal costs have remained low compared to alternatives, as in the US and China. Higher energy pricing encourages substitution by renewable fuels and energy efficiency. The same renewable technology would require more incentives in North America than Europe to remain economically neutral, since the latter generally has higher energy rates already.

4. *Material Exports & Energy Imports* – major exporting nations, such as China, have inherently high carbon intensities (as a percentage of GDP) as exports are often associated with heavy industry, which produces more emissions than light industry or service-based economies. The situation is worse with fossil fuel exporters; for example, the petroleum producing nations (e.g., OPEC) often have the highest emissions on a GDP basis. Importing nations don't have to account for the carbon emissions associated with production in other countries.

For meaningful reduction in GHG emissions, nations must accept controlled economic pain. However, the rules have to be "differentiated" to take into account the above (and other) factors to produce an equitable process. No one wants to play a game when the rules favor their opponents.

The treaty itself does not provide a working blueprint to resolve climate change, as it only establishes a target; each nation must set its own policies which establish the mechanisms to reduce GHG emissions. On a national level, the general types of responses are relatively few:

- Adapt our habitat to changing climatic conditions (such as rising sea levels, greater extremities in weather conditions, etc.) and/or our atmosphere to reduce warming by solar radiation management[15] and other geo-engineering options.
- Reduce emissions at the point source (e.g., using renewable energy sources and other low-carbon technologies that will cost more) and increase energy efficiency.
- Capture emissions so that they never enter the atmosphere.

The "ARC" needs to be constructed before the flood (literally) arrives.[16] There are also only three general mechanisms for politicians to promote such investments: regulate, tax and subsidize.[17] The climate change battle will truly be won when corporations adopt climate change mitigation into their business strategies. Some of the specific financial options and policy tools that governments can use to bring big business into the carbon game are presented below.

The plays

Legislation should set a series of emissions targets in a decade-stepwise fashion to give industry the time to approve and build the capital projects to reduce emissions. Once the targets have been set, government must build a support structure for encouraging cost-effective methods to achieve them. A national climate policy roadmap is developed based on the following principles:

1. progressive carbon emissions targets consistent with the long-term target
2. investor-risk mitigation
3. social equity for the nation, region, municipality and individual.

The latter point is pivotal, and requires some explanation. This means that such legislation will not cause material job losses or economic slowdown (i.e., some jobs may be lost and/or growth may slow but not to the extent of upsetting the overall economic health of the country). Government must ensure that the burden does not reside in one specific region. If one area is especially impacted by the regulations, ways to distribute the economic burden must be implemented, such as tax shifting whereby rebates are issued.

Legally binding targets are supported by policies that are: long-term, credible and clearly defined, with a guaranteed political framework including incentives and other commitments for industry. Climate policies must be impervious to political interference from changing political agendas, annual budgets and governments. The exception would be a "ratchet mechanism" to allow adjustments in future emissions targets if the climate change situation deteriorates faster than projected, or if government policies do not work as anticipated. Also, during periods of economic duress, government has the option to temporarily suspend requirements to reduce emissions.

Regulation must inevitably decrease the demand for fossil fuels. Fossil fuels are used for a simple reason; they are the least expensive fuel. Renewable energy sources technologies today, and possibly for a long time to come, cannot compete directly against fossil fuels in most jurisdictions. Therefore, mechanisms are required to direct us away from fossil fuels. Government policy must bridge the cost differential and encourage low-risk, low-carbon capital investments. Programs should include a combination of short-term incentives to get projects off the ground and long-term financial assistance. Governments must CRAFT policies to encourage carbon reductions:[18]

1. Carbon Price – government must help create a value for carbon
2. Renewable Energy Sources (RES) electricity rates set through a technology-neutral FIT program
3. Alternative Energy Development – promoting research, demonstration and commercialization of new RES technologies
4. Financing:
    i.   grants and other capital cost incentives
    ii. loans – higher-risk, interest-free, interest deferred or guarantees
    iii. "Eco-bonds" – to help raise funds, without using taxes, a green bond could be issued specifically to fight climate change; the issue would allow the public to show their support for such policies
5. Tax Structure – capital gains and credits; flexible depreciation and tax-loss carry forward schemes; income-trusts; venture/angel tax credits (innovation and productivity tax credits); flow through shares.

Another instrument is a government-supported financial institution, "The Climate Bank" (TCB), with the *raison d'être* to supply start-up capital for climate change projects. Such institutions already exist with the British Green Investment Bank and the ERP, Umwelt and related programs of the KfW Bank in Germany. The focus of the TCB would be to increase the availability of project capital through:

1. *Loan Guarantees* – TCB should provide risk sharing to domestic banks for loans. The loan risk evaluation system will include a factor for the "carbon value" of the project, which would be assumed by TCB through a loan guarantee to the main financial institution.
2. *Direct Investments* – interest-free debentures, which at the end of the period, TCB has the option to convert into equity, or into a real loan.
3. *Low-interest loans* and other financial instruments to reduce the overall cash costs.

As part of this initiative, a carbon price will have to be established, which businesses need in order to plan and justify their capital expenditures – carbon bucks must induce business to invest. A carbon tax, or emissions trading, would set the carbon price. A carbon tax establishes a definite price on carbon, which reduces the risk of forecasting the economic return of capital projects to reduce emissions. However, such a "tax"[19] seems unlikely, as in many jurisdictions the word "tax" is synonymous with political suicide. In fairness, "tax" is not the proper word; it is a fine, or penalty placed on emissions. If we must declare it as a tax, it should be grouped with the other "sin" taxes.

A simple example of such a policy incentive would be a vehicle[20] carbon tax/ credit based on gas mileage (or better still on carbon emissions). Vehicles above a standard are charged a floating surcharge and vehicles below receive a credit.[21] The tailpipe surcharge or credit is also a rating that would allow consumers to make better-informed decisions. Those who demand their gas guzzlers, then, must pay for them. The consumer has the option of improving the environment and saving money every day on fuel costs and sticker price, or buying the gas guzzler. A precautionary note on electric vehicles is that they are only carbon neutral if the generation of electricity is carbon neutral. Is there any point having an electric car if the electricity comes from coal (without carbon capture)?

Another example is an energy tax on electricity generation (a surcharge could be levied based on carbon intensity). Nothing spurs on energy efficiency better that an increase in electricity rates. A "green" tax on electricity that produces a marginal rise in costs will result in opposition to such policies. But let's look at an example of such a tax. Canada used 567 TWh of electricity in 2007. This country is blessed with some of the lowest power rates in the world at \$0.06 kWh. If this rate was increased by 10 percent (\$0.006 kWh), it would generate over \$3 billion per year. An average home uses 11,000 kWh per year, which results in a total cost increase of \$66 per year. Are taxpayers willing to pay \$66 per year more to initiate programs to fight climate change?

A carbon price can also be established by emissions trading, which would meet with far less opposition than a carbon tax,[22] and, although highly criticized in this book, could be applied provided nations do not manipulate the allowances for their own self-interest. Low caps are good for climate change but increase the cost to industry (and vice versa). Government must carefully balance the two and still achieve the long-term emissions target. Industry will always lobby for higher caps because of potential job losses and damage to the competitiveness of industry. The EU experience is that such industry claims are exaggerated.[23]

Also, we should consider other options[24] besides the traditional cap-and-trade model: cut-and-trade – the controversy over allowance allocation can be removed if replaced by credits for actual performance. In other words, only facilities that have reduced their GHG emissions can sell the excess carbon credits into the ETS. The process, then, is based on CO_2 accounting instead of forecasting. In a cut-and-trade system, the market translates ecological performance into economic rewards.

Winning the game

World governments have relied on business-as-usual policies to foster action against climate change, an approach which does not work. Tough decisions cannot be avoided; the most important of which is to pass legislation that directly limits emissions from coal-firing. The two major sources of emissions are electricity generation and transportation fuels. The transportation sector has two independent solutions: the replacement of fuels with ethanol, biodiesel and other renewable fuels and, second, electric vehicles. The latter, of course, depends on the carbon intensity of electricity generation. Consequently, decarbonizing electricity generation has the potential to reduce emissions from both sectors with the same investment.

Government can set a minimum renewable energy portfolio for utilities (quota systems), such as the Renewable Portfolio Standard in California, Renewables Obligation in the UK or the 2020 guidelines for the EU. Options for low-carbon sources for power generation include:

1. Carbon Capture and Sequestration (CCS: when CO_2 is scrubbed or captured out of the flue gases of the power facility). The IEA has concluded that:[25] "The only technology available to mitigate greenhouse gas (GHG) emissions from large-scale fossil fuel usage is carbon dioxide capture and storage (CCS)";
2. Nuclear power;
3. Renewable energy sources – bioenergy, hydro and related water technologies, geothermal, wind and solar: the most economic and proven RES must be advanced first.

When these technologies have been implemented, they must have *priority access to the grid (first-on, last-off basis)*. In other words, when less power is required, the first facilities to be taken offline will be fossil fuel-based generating stations; when more power is required, the low-carbon emitting stations will start up first. In this way, low-carbon electricity generation is maximized.

For many power generation facilities, we must keep on using fossil fuels for now. Nevertheless, responsible actions can still be undertaken. No new coal-fired power utilities can be built, or existing facilities modernized, without minimizing emissions. A simple short-term action is to replace coal with natural gas where it makes economic sense to do so. Natural gas produces over 40 percent less CO_2 emissions on an equal energy basis than coal.[26] Most of the GHG emissions reductions in the UK, which is second only to Germany in performance, were made by switching to natural gas, and the US has recently undertaken this transformation with the shale gas boom (US emissions in 2012 were 11 percent below the peak year of 2007).[27]

If there is any doubt about this focused approach, one only has to look at the emissions of the coal sector in the US. Among facilities, power plant emissions are the largest GHG emitters (Table 5.2). Of the 100 facilities that emit more than 7 Tg of CO_{2e} per year, 96 percent are power plants.[28] Power plants alone accounted for 2,324 Tg of CO_{2e}, which is about 40 percent of total US emissions.

The electricity sector focus allows for a surgical approach to GHG emissions. Foremost, all industry and households use electricity, so the economic pain is shared. Addressing this issue at the source point should mean that the direct capital burden on other industries could be greatly reduced. Last, the usual pitfalls

Table 5.2. Carbon statistics (GHG emissions) for the large facilities in the US: 2010.

	Total GHG (Tg CO_{2e})	No. of Facilities
Power plants	2,324	1,555
Refineries	183	145
Chemicals	175	541
Other industries	159	1,770
Landfills	117	1,199
Metals	99	265
Minerals	96	349
Pulp and paper	46	227
Government and commercial	15	173

Source: EPA, *Climate change, Greenhouse Gas Data.*

of industry competitiveness and job losses associated with restrictions on GHG emissions are less relevant in this sector, as utilities will not close because:

1.   cost increases will be diluted over all energy consumers
2.   regional supply sources are necessary to meet demand
3.   there is limited competition from outside the nation.

It is worth stressing, if we do not reduce carbon emissions from coal-fired utilities, we will be adapting to changing climatic conditions, instead of mitigating their occurrence. Without tackling the power sector, the climate change problem will never be resolved. Given the initiatives in Ontario, which were discussed earlier, we can label this strategy as, fittingly, the "ON solution." We must find ways to reduce CO_2 emissions without turning the lights off!

We are dealing with, arguably, the most complex political issue of our times, as prosperity and emissions are hopelessly intertwined in an ecological and economic Gordian knot.[29] An ancient oracle had prophesized that the one who untied the knot would become king; Alexander the Great failed to untie the knot, but he did cut through it with his sword. Simple solutions must be implemented; anything else will only bog the entire system down, as has been witnessed in the UNFCCC for the past 25 years. In this chapter, *simple* actions have been proposed to address our failure to even curtail GHG emissions; including:

1.   *Policy neutral rule* – this should be removed from the IPCC as politicians need better direction from the world's leading experts;
2.   *G3 negotiations* – need to simplify climate change negotiations for the establishment of carbon targets from 200 political entities to three, the US, China and the EU, which are the largest emitters;
3.   *Differentiated carbon targets* – carbon targets must not give a competitive advantage to one nation;
4.   *Carbon value* – carbon emissions must have a cost associated with them, through a carbon "sin" tax or a "cut-and-trade" market-based system; a

carbon value must motivate the market to adopt emissions as a metric of overall business performance;

5.   *Coal-fired utilities* – without directly tackling this specific sector, emissions cannot be brought under control.

In the future historians may praise the IPCC for its efforts but mourn the missed opportunity when governments did little to prevent the calamity now upon the world, and wonder why earlier societies were so complacent. They will likely quote, incredulously, the British House of Lords report which brushed aside the concern about future generations with a complaint about "costs": "Moreover, those costs will largely be borne by the current generations, while the benefits will accrue to generations yet to come, who are projected to be significantly wealthier and technologically more advanced."[30]

Climate change really is the Gordian knot of our times, requiring an "Alexandrian solution."

References[31]

Adams, B., and Luchsinger, G., UN, Climate Justice for a Changing Planet, 2009; http://www.un-ngls.org/IMG/pdf_climatejustice.pdf

Amin, A.T.M.N., UNEP, *Reducing Emissions from Private Cars*, October 2009; http://www.unep.ch/etb/publications/Green%20Economy/Reducing%20emissions/UNEP%20Reducing%20emissions%20from%20private%20cars.pdf

Ariely, D., *Predictably Irrational: The Hidden Forces that Shape Our Decisions*, HarperCollins, 2008; http://books.google.ca/books?id=ro7X8HRyuEIC&pg=PA305&dq=%22It+is+difficult+to+get+a+man+to+understand+something,+when+his+salary+depends+upon+his+not+understanding+it!%22&hl=en&sa=X&ei=RYQIT6I7pJyJApOErckJ&redir_esc=y#v=onepage&q=%22It%20is%20difficult%20to%20get%20a%20man%20to%20understand%20something%2C%20when%20his%20salary%20depends%20upon%20his%20not%20understanding%20it!%22&f=false

Bodnar, P., The Political Economy of Climate Change: Did States Defend Their Material Interests When Negotiating the Kyoto Protocol, *Proceedings 2004 Annual Meeting of the American Political Science Association*, September 2004; http://www.allacademic.com/meta/p_mla_apa_research_citation/0/5/9/9/5/p59951_index.html

Bolin, B., *A History of the Science and Politics of Climate Change*, Cambridge University Press, 2007

BP, *Statistical Review of World Energy*, 2013 (and earlier years); http://www.bp.com/en/global/corporate/about-bp/statistical-review-of-world-energy-2013.html

Cass, L.R., *The Failures of American and European Climate Policy*, SUNY, 2006; http://books.google.ca/books?id=tlrdtIz1w5wC&pg=PA33&lpg=PA33&dq=Those+who+think+we%E2%80%99re+powerless+to+do+anything+about+the+%E2%80%98greenhouse+effect%E2%80%99+are+forgetting+about+the+%E2%80%98White+House+effect.%E2%80%99&source=bl&ots=1STRiGZr62&sig=tb1_GiJSf1Ua5U2zuQ7iV5EKuSA&hl=en&ei=-m2nTNaVDIjCsAOistHuDA&sa=X&oi=book_result&ct=result&resnum=1&ved=0CBQQ6AEwAA#v=onepage&q=Those%20who%20think%20we%E2%80%99re%20powerless%20to%20do%20anything%20about%20the%20%E2%80%98greenhouse%20effect%E2%80%99%20are%20forgetting%20about%20the%20%E2%80%98White%20House%20effect.%E2%80%99&f=false

CDIAC (Carbon Dioxide Information Analysis Center) – Oak Ridge National Laboratories – US Department of Energy; data taken on May 12, 2013; (note that historic data is periodically updated by CDIAC); http://cdiac.ornl.gov/trends/emis/meth_reg.html

Cole, D.H., The 'Stern Review' and its Critics, October 2007; http://papers.ssrn.com/sol3/papers.cfm?abstract_id=989085

Dobriansky, P.J., and Turekian, V.C., Climate Change and Copenhagen: Many Paths Forward, *Survival* 51, 6, p. 21, November 2009; http://www.informaworld.com/smpp/content~content=a917062571~db=all; accessed November 2011

ECOFYS (for IEA), de Jager, D., and Rathmann, M., *Policy Instrument Design to Reduce Financing Costs in Renewable Energy Technology Projects*, October 2008; http://www.ecofys.com/files/files/report_policy_instrument_design_to_reduce_financing_costs_in_renewable_energy_technology_pro.pdf

EIA, *Voluntary Reporting of Greenhouse Gases Program*, January 31, 2011; http://www.eia.doe.gov/oiaf/1605/coefficients.html

EPA, *2010 Greenhouse Gas Emissions Data from Large Facilities Now Available*, News Releases by Date, January 11, 2012; http://yosemite.epa.gov/opa/admpress.nsf/0/8890DDDC08B1B82785257982005CCAC; accessed January 2012

EPA, *Climate Change, Greenhouse Gas Data*; http://epa.gov/climatechange/emissions/ghgdata/; accessed January 2012

Freed, J., and Hodas, S., Putting a Price on Success: The Case for Pricing Carbon, Clean Energy, May 2010; http://thirdway.org/publications/291

Grubb, M., Brewer, T.L., Sato, M., Heilmayr, R., and Fazekas, D., Climate Policy and Industrial Competitiveness: Ten Insights from Europe on the EU Emissions Trading System, August 2009; http://www.climatestrategies.org/component/reports/category/61/204.html

Hannam, P., Former UN Official Says Climate Report will Shock Nations into Action, *WAtoday*, November 7, 2012; http://www.watoday.com.au/environment/climate-change/former-un-official-says-climate-report-will-shock-nations-into-action-20121106–28w5c.html; accessed November 2012

Helm, D., Climate Change Policy: Why Has So Little Has Been Achieved; *Oxford Review of Economic Policy*, 24 (20), 211, 2008; http://www.kysq.org/docs/211.pdf

Helm, D., To Slow Warming, Carbon Tax, *New York Times*, November 11, 2012; http://www.nytimes.com/2012/11/12/opinion/on-climate-change-the-us-is-doing-better-than-europe.html?partner=rssnyt&emc=rss&_r=0; accessed November 2012

Hepburn, C., Carbon Trading: A Review of the Kyoto Mechanisms, *Annual Review of Environment and Resources*, 32, 375, 2007; http://www.eci.ox.ac.uk/~dliverma/articles/Hepburn%20carbon%20trading%20ARER.pdf; accessed November 2011

IEA, *CCS*, 2005; http://www.ipcc.ch/publications_and_data/publications_and_data_reports.shtml#.UcnU3tj8-V4

InterAcademy Council, *Climate Change Assessments, Review of the Processes and Procedures of the IPCC*, 2010; http://reviewipcc.interacademycouncil.net/report/Climate%20Change%20Assessments,%20Review%20of%20the%20Processes%20&%20Procedures%20of%20the%20IPCC.pdf

IPCC, *First Assessment Report, Working Group III: The IPCC Response Strategies*, 1990

IPCC, *Report of the First Session of WMO/UNEP Intergovernmental Panel on Climate Change*,[32] Geneva, November 9–11, 1988

IPCC, *Report of the 33rd Session of the IPCC, Decisions Taken with Respect to the Review of IPCC Processes and Procedures, Communications Strategy*, May 10–13, 2011

Meixian, L., Rebates for Low Carbon Emissions Cars from 2013: LYA, *The Business Times*, November 28, 2012; http://www.businesstimes.com.sg/breaking-news/singapore/rebates-low-carbon-emissions-cars-2013-lta-20121128; accessed November 2012

Moss, R.H., The IPCC: Policy Relevant (Not Driven) Scientific Assessment, *Global Environmental Change* 5 (3), 171, 1995

Nisbet, M.C., Communicating Climate Change, *Environment* 51 (2), 14, 2009

Nordhaus, W., The Challenge of Global Warming: Economic Models and Environmental Policies, April 2007; http://qed.econ.queensu.ca/pub/faculty/lloyd-ellis/econ835/readings/dice2007.pdf; accessed November 2011

OECD, *Taxation, Innovation and the Environment*, October 2010; http://www.oecd.org/document/6/0,3343,en_2649_34295_46091974_1_1_1_1,00.html

O'Neill, S.J., Hulme, M., Turnpenny, J., and Screen, J.A., Disciplines, Geography, and Gender in the Framing of Climate Change, *Bulletin of the American Meteorological Society* (August), 997, 2010

Sarewitz, D., Curing Climate Backlash, *Nature* 464, 28, 2010

Shackley, S., The Intergovernmental Panel on Climate Change: Consensual Knowledge and Global Politics, *Global Environmental Change*, 7 (1), 77, 1997.

Stern, N., Stern Review on the Economics of Climate Change, 2006; http://webarchive.nationalarchives.gov.uk/+/http://www.hm-treasury.gov.uk/sternreview_index.htm

The Royal Society, Environmental Defense Fund, and Academy of Sciences for the Developing World, *Solar Radiation Management: Governance of Research*, December 1, 2011; http://royalsociety.org/policy/projects/solar-radiation-governance/

UK House of Lords, Select Committee on Economic Affairs, *The Economics of Climate Change*, July 6, 2005; http://www.publications.parliament.uk/pa/ld200506/ldselect/ldeconaf/12/12i.pdf

Weber, E.U., and Stern, P.C., Public Understanding of Climate Change in the United States, *American Psychologist*, 66 (4), 315, 2011

Notes

1 Weber and Stern, 2011; also see Nisbet, 2009, p. 22.
2 Sarewitz, 2010.
3 IPCC, 1990, p. xxvi; for later examples see, InterAcademy Council, 2010, pp. xv, 54; Shackley, 1997, p. 79; IPCC, May 10–13, 2011, pp. 4–5.
4 Nisbet, 2009, p. 18; O'Neill *et al.*, 2010: and references therein; Bolin, 2007, pp. 211–13.
5 See for example, InterAcademy Council, 2010, pp. xv, 54.
6 An early defense of this policy is presented by Moss, 1995, p. 172.
7 IPCC, November 9–11, 1998, p. 2.
8 Stern, 2006, p. ii; the report presents a detailed overview of the economics of climate change; for an analysis of the report see Cole, 2007; also see Helm, 2008, pp. 224ff; for another economic report see Nordhaus, April 2007.
9 Hannam, November 7, 2012.
10 The availability of other forums to tackle climate change appears in Dobriansky and Turekian, November 2009.
11 Stern, 2006, p. 35.
12 See, for example, Bolin, 2007, p. 245.
13 For a discussion of the general differences between nations see Bodnar, September 2004. Political structure also has an impact; see Cass, 2006, pp. 15–17.
14 Ariely, 2008, p. 305.
15 For a review of this controversial area, see The Royal Society, December 1, 2011.
16 Right now we seem to be on another vessel, the *Titanic*, since our grand atmospheric ship is about to hit a "dry" iceberg. For a review of the social consequences of climate change, especially in the more vulnerable developing countries, see Adams and Luchsinger, 2009.

17 A review of various policy initiatives is presented in ECOFYS, de Jager and Rathmann, October 2008.

18 The OECD has an excellent website on "Financing Climate Change"; http://www.oecd. org/document/16/0,3343,en_2649_34361_43577616_1_1_1_1,00.html

19 See Weber and Stern, 2011, p. 322; a recent review has been published on environmental taxes, see OECD, October 2010; also see Hepburn, 2007, p. 378; Helm, November 11, 2012; and Freed and Hoads, May 2010. For the issues in the carbon tax debate in Germany and Europe, see Cass, 2006, pp. 64–5, 70. On July 1, 2012, Australia introduced a carbon tax.

20 The United Nations Environment Programme has recently issued a report on private automobiles and emissions (see Amin, October 2009).

21 Such a program was being initiated in Singapore in 2013; Meixian, November 28, 2012.

22 Grubb *et al.*, August 2009, pp. 3, 15; provides an overview of designing a new ETS system based on the EU experience; see Stern, 2006, 15.4, pp.330+.

23 Grubb *et al.*, August 2009, pp. 8, 14.

24 For a review of ETS options, see Grubb *et al.*, August 2009, p. 19; Hepburn, 2007, pp. 383–4.

25 IEA, 2005.

26 EIA, *Voluntary Reporting of Greenhouse Gases Program*.

27 BP, 2013.

28 EPA, January 11, 2012.

29 According to legend, King Midas of Phrygia, the son of Gordias, had created the knot, which became a symbol of their rule. Alexander was the first one to "think outside the box."

30 UK House of Lords, July 6, 2005, p. 43.

31 All web-based references were confirmed (accessed) in June 2013, unless otherwise stated.

32 IPCC Plenary Session minutes and supporting documents can be found at http://www. ipcc.ch/meeting_documentation/meeting_documentation_ipcc_sessions_and_ipcc_ wgs_sessions.shtml

Index

For Product Safety Concerns and Information please contact our EU
representative GPSR@taylorandfrancis.com
Taylor & Francis Verlag GmbH, Kaufingerstraße 24, 80331 München, Germany

www.ingramcontent.com/pod-product-compliance
Lightning Source LLC
Chambersburg PA
CBHW050423280326
41932CB00013BA/1973

9 781138 186880